A human being is a part of the whole, called by us, "Universe," a part limited in time and space. He experiences himself, his thoughts and feelings as something separated from the rest ... a kind of optical delusion of his consciousness. This delusion is a kind of prison for us, restricting us to our personal desires and to affection for a few persons nearest to us. Our task must be to free ourselves from this prison by widening our circle of compassion to embrace all living creatures and the whole of nature in its beauty. Nobody is able to achieve this completely, but the striving for such achievement is in itself a part of the liberation and a foundation for inner security.
- Albert Einstein

UNIVERSE CITY

We cannot always build the future for our youth, but we can build our youth for the future.

Franklin D. Roosevelt

CHAPTER 1: JAIOBIAN (pronunciation: Hi-oh-bee-awn)

The ridges pierced through the cloudless blue sky; as yet another day of unusually high temperatures scorched the countryside. "No rain again today." She thought to herself, shaking her head, as she took in the thinning foliage lining the canyon and what was once a ravaging river receding in a dying land.

He cleared his throat, then shouted, "She didn't remember that it's your birthday...did she?!" The uncomforting question rumbled in her ears, slightly echoing its affirmation and reminder, against the rocky walls.

Jaiobian sat down on a smooth area of a boulder at the edge of the diminishing flow, looking down at the dirty water as it attempted to push varied pieces of trash through the other large rocks. Some pinching so tightly that they formed small pools in which the trash accumulated.

She shrugged and discernibly whispered, "I guess not." She released her lungs held breath in a soundless sigh, her shoulders lowering.   Nine years old this day and life had cruelly taken its toll. Her frail frame was a reflection of neglect.

"Seriously, Jai, she never remembered, uh, remembers much of anything, period. She however, seems to never forget to collect the money that you get from the tourists. I've seen it over and over. It's like you owe it to her, so she can shoot up and forget that she or maybe you ever existed." Mister Richards bore witness to some of this child's horrors. His words were harsh but factual, and for reasons yet to come, he attempted to bring the reality of the situation to a necessary forefront.

"Stop it already! She has a problem, okay, I get it! You don't think... I know that?" Her dark eyes took on a smidgeon of the sinister. "But at least she didn't cut me out of her stomach, at least she had me, you know?" Jaiobian stood erect her shoulders back, her feet slightly apart challenging Mister Richards in an attempt of defiance and a wisp of dignity.

"Is that what she reminds you about if you complain?"

Jaiobian's eyes saddened briefly, "Not all the time, no." She fidgeted, attempting to brush her emotional pain aside. However her minds' blockades couldn't hold.

Mister Richards persisted, "But some of the time?"

Jaiobian whispered, "Yeah, sometimes, I guess."   Her mouth twisted from the memories that entered her thoughts through the barriers that had been built to protect her over time. Her anger increased. In an attempt to keep her lips from trembling and losing control she glanced downward and continued to look at the pieces of debris. She focused in hope of finding something that she might be able to sell. She knew that if she cried, it would only make her feel worse, and she might lash out violently at Mister Richards; one of few adults that she had grown to trust. Besides, there was no one available to give her pity, except maybe for Mister Richards. She bit down on her bottom lip and tried to maintain her focus on the colorful array of disregarded candy wrappers, and varied plastics. She shook her head in disappointment of any possible treasure amongst the items, and in attempt to clear her mind of the unusual conversation.

Mister Richards, under normal circumstances, would not have been so crude. No, this time, events had come to a head and he was almost forced to initiate his plan that had ripened over the past year. In this part of the world, there were no organizations that could help this child except perhaps the one that he represented. Although once she did receive a blanket and a bar of soap when she agreed to have her photo taken. "You're an incredible human being, Jai, and I have an offer for you?"

Jaiobian turned her head slightly as she glanced at Mister Richards suspiciously and spoke firmly, "I don't do nothing for nobody. Get it?" Her voice carried as that of a tongue tied child, but she was not.

"No, no, no, it's not that sort of offer. This is a chance of a lifetime. It's your ticket to freedom and happiness."

"Huh? Ticket?" Jaiobian mumbled as she attempted to move back further on the boulder. Her posture and little clothing over her small frame made her appear as if a large frog had taken residence.

"Yup, an actual ticket to paradise." Mister Richards smiled, his arms opened in anticipation.

Jaiobian stood yet again and turned to leave. Her trust, if any, had thinned past suspicion and now progressed toward protection. She sought to defend whatever amount of significance she had created for her own sanity. It wasn't Mister Richards' fault, as it was instead the many promises never kept, the many lies of a generality of people; nameless, faceless, and yet affective and thorough in their degradation.

"Where are you going?"

"I need money." Jaiobian stated as she turned her head to look in Mister Richards' direction. Then with a look of realization and quite an amount of anticipation, she explained: "Maybe I can get enough to buy a cupcake! A real pretty bright pink one, with lots of sprinkles and a candy rose on top. I saw one in the window of Antonio's Bakery. I gotta go, bye!" Jaiobian jumped off the boulder and onto the dirt embankment. She walked slowly, just in case Mister Richards reached for his wallet."

"Jai, wait! I'll get you that cupcake and we can sit at the bakery and talk a little more."

Sure! Okay." Jaiobian wasn't about to let such a deal slip by.

She was cunning. Her eyes glistened with excitement. She knew how to play the game, how to manipulate for money. She'd remain cautious, walk behind not ahead, and if Mister Richards tried anything that she didn't agree to, she had back up. She had her gang of nine other children of various ages.

Most were small in stature, but in such a number, they were feared. Bobby or "Blade" alone was ruthless with the knife. His nickname given to him by the rest of the gang for what seemed obvious. His favorite pastime was sitting on the boulders at that rivers' stream and sharpening the blades of his assemblage against the rocks, sliding them ever so gently, then holding them up to the sunlight, turning the handle and watching them sparkle as the light caught the razor sharp edges. He had an impressive collection even for a machirologist. He'd challenge anyone in whatever knife game they chose in winner take all matches, and so far, he had never lost.

Jose or "Twig" as they called him, was tall and skinny, over six feet tall by his eleventh birthday. He could swing a branch like a pro on a baseball field. He once broke a man's arm and then continued to beat him until the rest of the gang managed to pull him away. His target was the arm that was clutching on to Carla, better known as "Craze", who was about the fastest pick-pocket around and could empty every pocket on a magician in less than a minute. The others had their varied motley skills. Twig's only thought was that Craze was in trouble, and he moved in quickly in order to free her. It turned out that Craze didn't steal anything. What Twig was in fact sure of was that the man was in fact attempting to take Craze. That was never going to happen while he was around. He watched his step-father beat his little sister to death. He tried to help her, but he was much shorter and just as skinny back then. His step-dad sent him flying across the room. He woke up in a local clinic. His mother, sister and step-father were all dead. Before the police could get to his step-dad, he had committed suicide. Twig appeared to be on a mission of sorts, to protect anyone that was being physically hurt. He'd occasionally walk onto the near-by school yards if he noticed someone being bullied.

Blade swore to kill the man next time he saw him, and Twig swore to help, but the guy must have left the area, because no one ever saw him again. Some wondered if Blade and Twig had anything to do with his disappearance, but no one ever asked.

Walking on the rough road to Antonio's was tough on not only Mister Richards' sandals but his feet. He looked down at Jaiobian's spread out toes. Her callus ran up the sides of her feet with the skin on top resembling leather. It was as though nature provided shoes through adaptation as these were apparently much better than his sandals. Jaiobian appeared to be proving such by intentionally skipping alongside his slow pace, in order not to get too far ahead. On occasion he'd stop to remove a pebble while sounding off his discomfort. Jaiobian would giggle and then cross her arms, and stop. "Humph, not again" Jaiobian muttered, by now a bit perturbed. Mister Richards had briefly attempted to remove his shoddy slip shoes, but almost immediately returned them to his feet once a sharp edge of a rock met with skin.

"You should try to walk without those Mister Richards and get used to it, then you can have feet like me." Jaiobian smiled proudly.

Mister Richards nodded; "You do have amazing feet, Jai." Mister Richards stopped once again and took off a sandal removing another pebble that had lodged between his toes, then mumbled, "Indeed. I could sure use feet like you have. I've gone through so many sandals." He shook the sandal and returned it to his foot. "They wear out pretty quick around here." He looked around at what was called roads but were in actuality cut paths through the harsh vegetation. The bull dozers left trails of rocks and very little dirt.

Just over the ridge, no trees remained. Their main purpose was in clearing the forests. Vehicles could be heard hitting and sometimes dragging their bottom carriages against the larger rocks.

While waiting she wiggled her toes around and pondered her fortune in having such feet.

Antonio's was the only bakery in town and soon they would walk through the undersized entrance that allowed one person to enter at a time. Antonio couldn't afford to install an actual turn style so he innovated and purposely built the door smaller than average, in an attempt to reduce crime. However, what seemed like a bright idea at the time would prove for the more robust sized customer to be counterproductive. The makings of another door was in its construction, following several complaints, and his new help should make it unnecessary.

The stare of the woman behind the counter was one of directed contempt for Jaiobian. Defiantly Jaiobian disregarded the glare, and walked through the door holding her head high, her small frame moved easily between the support beams. She pretended that the door was made that way for children. Perhaps specifically for the special children as petite as she was. Her mind imagined a modest yet grand entrance in a world made of sugar.

Antonio was rarely at his bakery as it seemed to run more efficiently when the manager that he had recently hired wasn't being critically watched, or at least he had thought. What he wasn't aware of is how cruel that manager could be, he only knew that product was selling and he no longer was robbed and now was making somewhat of a decent profit. The manager, his manager, also ran most of the small businesses including the children's detention center, a front for a children's' sex trade operation. No one appeared to notice the strange and steady disappearances of the many homeless children in the area. Perhaps the inattention was more of an indifference to the steadily growing impoverishment. However such apathy allowed the manager to move and manipulate children with ease.

Mister Richards was aware that something was suspiciously wrong. His organization had attempted to talk to the children in the detention center several times, but was continuously turned down. The local government officials supported the centers' decision fully, claiming that it was a great benefit to the area, which only enhanced Mister Richards' and his teams' suspicions. It didn't make sense that children being locked up would be considered a "benefit." They had sent several complaints and even hired attorneys to circumvent the barriers toward entry into the facility, but so far nothing seemed to work. Mister Richards' organization offered to help with the children, but his organization was instead threatened with deportation. Mister Richards was forced to pull back in order to complete his mission of saving as many children as was possible.

Meanwhile, when Antonio did appear, and on those almost rare moments in which the children of the forests cautiously checked daily, he would place a plate of cookies right outside that door. The children reminded him of birds swooping in as they would grab what they could and then run away quickly before Antonio came through that door with his broom in hand screaming and acting as if they were robbing him. Jaiobian and some of the others had caught him laughing on a few occasions. It eventually became a game to which the children would sometimes pull at Antonio's apron simply to provoke him to chase them, as they ran away screaming and laughing. Not seeing that plate of cookies always brought audible sighs, as the children would slowly walk back into the forest.

Jaiobian's gang never robbed Antonio's business, but apparently other neighboring gangs had more than once. Antonio wasn't sure of who or why. He felt that there was no other alternative than to take on a separate bakery operator other than himself. After all, he had children of his own, seven in total, four boys and three girls. His heart was empathically pulled, yet removed, thus ambivalently strained and cognitively dissonant. He did what little he could perhaps for his own relief of guilt and fear. For the little moments of laughter and the confirmation that these were in fact children.

It was too easy for him to imagine that if something were to happen to him what might happen to his own children. His wife was ill, many were ill in this part of the world, and his own health wasn't the best. Maybe it was the fracking, or the war, no one knew for sure. What was left of the river was filthy, and water ran brown at times through the pipes, and smelled heavily of chlorine, and burned when lit. He was one of the few that could afford a private filtration system. But even that was proving futile. His eldest was only eleven. He wanted to help the children of the streets, but he feared them too much, mostly because there were others that depended on him. Death and or debt was not an option, and at times he felt trapped in limbo.

Today, however, neither Antonio nor the manager was there, just the baker woman who had been mugged more than once by the local gangs. She shook her head as Jaiobian walked into the bakery with Mister Richards behind her. Jaiobian felt a sense of pride as she knew that she would not be kicked out, nor commanded to pay first and wait outside the door for her cupcake. Instead she would in fact be allowed to sit at a table because everyone in town was familiar with Mister Richards' funded research in regard to children. And not only that, but he bathed and dressed nice, and didn't have to scratch himself all the time as she did. She even once saw him in a suit! No one ever dressed in suits except for the few politicians and lawyers, the people with lots of money. She skipped to the table nearest the window, and plopped down on the chair, smiling. She briefly thought that this is what it must feel like to have a dad, as she glanced toward Mister Richards.

Mister Richards entered by slightly angling his shoulders and bending his neck in order that his six foot two inch frame fit. It wasn't necessary, but he wasn't sure. He looked over to where Jaiobian had chosen to sit and thought to himself how amazing it was that these children held onto a sense of dignity. It was almost as if they had a claim to an innate right, the right to exist. Jaiobian looked down and then toward the window once Mister Richards' eyes met hers'.

The cupcake tasted like a pink cloud as it glided past Jaiobian's parched lips and then slowly over her tongue. For a moment, her lips turned a bright pink; she hadn't before, but now she noticed her reflection in the window. The color on her lips brought her image to her attention. She moved her matted hair off her face and smiled slightly, the image wasn't much to her liking. Smacking her lips, she attempted to make the pink frosting look like lipstick. She then pulled back most of her hair to reveal her darkened skin and imagined being a princess. "But Your Majesty...what shall we do with all the other cupcakes that are not to your liking?" "Give them to the crows my dear fellow." Her lips moved in silence.

Mister Richards' eyes crinkled, "Huh? I can't hear you?"

Jaiobian noticed his stare, "Oh, I wasn't saying anything." She focused again on her cupcake and took another bite, then with mouth full and chomping openly, she picked up the glass of milk. He even bought her a glass of cold milk! Cold fresh milk in a clean glass! Not warm like the glasses of milk that she waited for when the tourists would leave their tables.

Mister Gonzales the owner of the only restaurant in town would allow their gang to finish the food on the tourist's plates. Every day, they would wait patiently in the bushes as to not be seen by the tourists and possibly upset them; hoping that they weren't so hungry. They would listen for the signal; the dinner bell that Mister Gonzales would ring and they would run as fast as they could for the plates of leftovers. Nothing was ever left behind, not even a crumb for the birds. Jaiobian however, would never forget to put what she could in her pockets and throw what she could afford to go without, in the parking lot for the crows.

She adored the crows as they were perhaps the most misunderstood of all the birds. They were smart and made her laugh with their antics. They too had to figure out ways to survive and they seemed unwanted because they couldn't sing sweet songs, and weren't considered amongst the pretty birds, the accepted birds like the various parrots in what was left of the forest. To Jaiobian, they, were very much Jaiobian. So much so, that, by agreement, their gang was called the "Crows."

Mister Richards began his delivery. "Jai, I've now known you for a full year, and, well, I've done quite a bit of research on you."

"Huh? Research? That's what I hear about you Mister Richards." She nodded and continued with what appeared to be an interview about Mister Richards. "But, tell me why do you do research? Like you maybe write a book?"

"Sort of, but not quite, Jai. I've been doing what's called a documentary about the Crows."

"Huh? Oh, I know what that is...like when some come and say that they want photos and no pay us. So, okay, you take photos, but we need money, no blankets."

"Well, no, not exactly Jaiobian. We don't want photos. I'm here in search of children that need our help, and that we feel may make a difference one day to this world. Uh, how do I explain...hmm?" Mister Richards paused in order to gather his thoughts. "Okay, well, there's a group of very rich people that have hired me to go around the world selecting children that have no parents and would benefit from these people's help, and therefore possibly benefit the world."

"That's what you said the first time, Mister Richards." Jaiobian grinned. Mister Richards nodded from his slight embarrassment at not being able to colloquially explain sufficiently to this child. Jaiobian however caught on very quickly, and she continued, "But, I have a mom."

"I know that, Jai. But, you still qualify."

"Qualify?"

"You are what they are looking for, as are your friends."

"For what?"

Mister Richards attempted to break it down to more understandable jargon, "They want to help you by giving you food whenever you want, and a place to live with guardians that will love you and take care of you. Where you can go to school and learn to read and write and be whatever and whomever that you dream of being."

"How can I be what I dream of being?" Jaiobian giggled, as Mister Richards certainly didn't know of her high aspirations of being royalty. She corrected her posture, and sat upright proudly, her hair so matted and tangled that when she pushed and pulled it back again it held stiffly in place as if being blown. Her face dirty, except where the milk had cleared the area around her mouth, leaving a slight white hue over her top lip.

"You can be whatever you put your mind to, Jaiobian, you'll see. (He paused) In fact, due to the data that I've acquired about the Crows, The Coalition is very interested in you particularly."

"Huh...why?"

"Well, Jai...we're relatively sure that you're what's called a "natural leader." Mister Richards gestured the quotation marks with his fingers.   "It's rare to find such traits in someone so young. Or, at least we think that it is. You're what The Coalition is looking for...in order to hopefully steer your natural abilities toward the benefit of others and the world. See, we can prove to the world that every child can be if need be, saved. Saved from detriment to not only the world, but themselves."

Jaiobian stared at Mister Richards. Mister Richards assumed that she wasn't sure what he was trying to relay. Jaiobian however was calculating the Crows possible benefit from this exchange.

"Well, I can see that this may be a bit confusing, so I'll leave it there, and once we get things going, you should begin to understand."

Jaiobian shrugged her shoulders, "What do we have to do?"

"Nothing. You and your friends will get on an airplane and fly to Universe City."

Jaiobian shook her head, "Whoa, get on an airplane? No, no, no way, I'm not leaving here. You got it wrong, Mister Richards, I'm not interested. Uh, unless?"

"Unless? Remember, you won't be alone, Jai. All your friends are coming too and besides, you can leave Universe City anytime that you want."

"Any time? Where is it? How far?"

"Yes, any time. If you get off the airplane and want to come back here, then you can turn right around and get on the next plane and leave any time to come back. It's a little over four hours away by airplane, and it's on its own island, but anytime you want to leave, you'll be flown back. Okay? I promise. How's that?"

Jaiobian looked into Mister Richards eyes. "Hmm. I'll think about it, with the Crows."

Mister Richards nodded and continued. "Before I approached you, Jaiobian, I had talked to all of them, and they all appear to want to go, but they want to know if it's okay with you."

Although young amongst some of the children in her makeshift gang called "Crows", they all looked to her for leadership and the final decision in anything that they attempted. She seemed to be wise beyond her years. The natural leader as it was called that persistently appeared to know how to keep them all safe and surviving. Besides, she was the one that made the deal with Mister Gonzales. He would allow them to eat the leftovers and they all agreed to never upset the tourists nor rob them or him.

Her talent appeared to be the gift of gab. She could talk a convincing tale on just about anything. And her abilities toward reasoning and negotiations were extraordinary. She delayed any agreements in order to push the deal further.

"Mister Richards?"

"Yes, what is it, Jaiobian?"

"How much?"

"You want me to pay you?"

"Yes." Jaiobian nodded, rubbing her chin in thought. "I say, one hundred for each of us with paper that says we can leave, any time."

"That sounds fair. I think that I can arrange that."

Jaiobian smiled largely. A hundred dollars was a significant amount of money. Multiplying by the ten Crows, wow, she wasn't sure how much that was, but she was sure that it was a great deal of money, and she was excited to close the agreement for everyone. "Okay, you got yourself a deal Mister Richards. You bring the money and the paper and the tickets tomorrow." Jaiobian spit slightly in her palm and then extended her handshake.

Mister Richards reciprocated by slightly spitting in his palm, smiled and shook hands on the deal. "I'll be here tomorrow with everything, Jai."

The following day, Mister Richards was at the bakery early, realizing that he never suggested a time. He sighed resting his chin on the palms of his hands as his elbows balanced on the table. He sat there waiting for over three hours checking his watch every ten to twenty minutes, regretting that he didn't bring anything to read; then rationalizing that this was a good lesson for him and that he would make sure never to rush these events again.

The woman behind the counter gestured occasionally as to whether he wanted a refill of his coffee. He waived off the offer each time; and then to his relief he counted all ten children as they approached the bakery. The woman attempted to chase them away with a broom. Mister Richards moved toward her quickly and bumped the top of his head on the door. Rubbing his head with his left hand he mumbled "ah shit!" then he grabbed part of the broom handle, and successfully removed the weapon. The embarrassed woman walked back behind the counter.

He stood on the small patio of the bakery and announced; "Alright Crows, I have your money and papers stating that you can leave at any time and here's your tickets." Mister Richards waved the tickets in the air.

The children's eyes widened as they immediately ran up to him attempting to instead grab at the money.

"Whoa, calm down, I have enough for all of you!" He chuckled, as he handed a hundred dollar bill to each one, counting to be sure that it was ten hundred dollars, yup, a thousand in total. "I'll be holding on to our tickets, by the way." Mister Richards put the tickets in his safari styled vest pocket, and then continued to hand out the last three bills.

Jaiobian waited patiently and wiped her hands with her dirty shirt, then took the last one of the hundreds. She then ran off.

"Jaiobian! We had a deal!" shouted Mister Richards.

She turned briefly; "I have to do something, be right back!"

Jaiobian ran as fast as she could to the small dirt floored shack. How the walls still stood was a mystery. She had spent her birthday with the Crows. In a hollowed out tree that they called their nest. She never bothered her mother when her mother was "working", which meant that when there was a man or men around, she was not to enter.

There appeared to be no one there. She walked in slowly, peering from side to side. Her mother lay on a makeshift bed of old blankets. No one else was next to her. Jaiobian approached her relieved. Her mother's arms were riddled with needle holes. Jaiobian's eyes fixated on the many bruises that lined her mother's arms, making her skin darker than usual. Jaiobian wasn't sure if her mother was alive, but that wasn't unusual. Her mother's breaths were hardly audible and her chest never appeared to rise. The drugs seemed to do that to her. She lied there as if in the stages of rigor-mortis. Once or twice, Jaiobian was forced to get the neighbors to wake her, but that proved nearly fatal as her mother never held back punishing her, in particular if Jaiobian had no money. Jaiobian debated whether to wake her this time. She should after all be happy that Jaiobian had such a bounty. Shouldn't she? Should she tell her mother that she would be right back once she took an airplane ride? Uh, no, that probably wouldn't be such a great idea. Instead Jaiobian pulled up the tattered blanket over her mother's shoulders and placed the hundred dollar bill on her chest. She then ran out the entrance that held no door.

Mister Richards was no longer at the bakery and neither were the other children. When Jaiobian questioned the baker woman, she shrugged her shoulders and offered no help.

Jaiobian ran down the street looking in every shop as she screamed; "Mister Richards! Mister Richards!" She finally came across Micah, a little boy that she envied for seemingly having the most perfect family. His parents would both hold Micah's hands as they walked on the sidewalk of the town talking and laughing and enjoying whatever Micah imparted.

Micah attended the nearby private school. Where on occasion, Jaiobian found herself prying through its gates, squinting and attempting to read the teacher's lips as they walked around in their classrooms apparently instructing something of great importance.

Jaiobian's mother had once walked her to the nearby public school, telling her to go in and register. But Jaiobian was seven years old at the time and had no idea of what to do, and many of the children stopped to stare with looks of distain. She wondered...why? Sure, she was in her usual rags, but she bathed the day before in the stream. She certainly didn't reek with odor as she usually did, or at least she thought that she didn't. Her mother did manage to get up that morning and cut off most of her hair. Maybe that was it? She looked a bit like a boy. Maybe the other children just weren't sure? She stood erect, swallowed, told herself that she could do this although she was terrified, and then she opened her mouth slowly and attempted to ask for help. "Ewe, go away!" The boy shouted, reacting to her by moving away and laughing, his friends joining in. The sense of disparity came crashing toward her. Every ounce of dignity removed as the feelings were overwhelmingly that she was in a place that she was not only not welcomed, but didn't belong. She ran, ran fast, ran out the doors, and into the forest, crying, and vowing to never again attempt such a task.

Her mother had asked her how it went in school that day. She responded that they didn't allow her to go. Her mother shrugged, "Oh well, you tried." Jaiobian wasn't sure if that comment was meant for her or her mother.

"Have you seen any of my friends, Micah?" Jaiobian asked between breaths.

"Yeah, they went that way with Mister Richards." Micah pointed. His parents nodded in agreement and appeared concerned. Everyone was aware of the disappearances of children in this area, but they were the homeless, and didn't appear to draw much empathy unless as in Jaiobian's case...many knew her and her mother, and some claimed to know whom her father might have been.

"Okay, thanks!" Jaiobian ran in the direction.

Down past the town was a converted fitness center that now served as a bath house for the town's poor. Jaiobian and her gang had never used the facility mostly because of the reports of violence that sometimes happened amongst the desperate. The river served their needs, although the river wasn't much of a river anymore and had been declared unsafe.

She slowly, very slowly entered through one of the large double glass doors as the other was locked into position and held numerous cracks. Her defense mechanisms were on full, as her fear was that… her family, the Crows, had been led to their slaughter. Maybe Mister Richards wasn't as nice as he appeared. Maybe he was like most of the other adults that she had come to know through their various cruelties.

She could hear several people talking and recognized some of the voices. At first she thought that she had heard Twig and then she was sure that she heard Blade and eventually Craze. There was laughter. Yes, some of the children were in fact laughing. She relaxed a little and walked in to see Mister Richards and a few other adults putting out new clothes and towels for the group.

"There you are Jaiobian." Mister Richards smiled and gestured for her to come in. "My staff has prepared all the necessary equipment before we can leave."

Jaiobian was a little winded and wiped her forehead with the back of her hand as she looked over toward two boxes; one had been opened to reveal bottles containing some sort of liquid which were being opened by a female staff member.

As she struggled slightly to open the twist top on a bottle, she noticed Jaiobian's stare and responded; "Oh, Hi. I'm Doctor Wilson, you can call me Tricia, (Doctor Wilson smiled) uh, oh, this is just lice remover and this other box is body wash and sanitizer. You girls will come with me and the other female aides and the boys will go with Doctor Richards and the male doctors, umm, aides, into the other showers."

"Doctor?" …"Richards?" Jaiobian's forehead wrinkled. She then continued, waving her arms and pointing to the boxes. "Oh no, no, all this stuff wasn't in the deal." Jaiobian turned to walk out.

The youngest member of her gang, a little girl named "Sheba" the name given to her by Jaiobian because she was roughly, by appearances, three years old when they found her walking the streets, and when asked what her name was, she would repeat "she bad." Sheba was now nearly seven years old and she pleaded with Jaiobian to let her play in the showers. "Please… please… please, Jai, please!" Her eyes were wide as the night owl as she tugged and bounced pulling on Jaiobian's torn shirt.

"Okay… okay, you're going to rip my shirt some more, stop it, ugh, alright, we can play for a little while." Jaiobian rolled her eyes and sighed in frustration.

"Yea!" screamed Sheba as she hugged Jaiobian.

Jaiobian looked sternly at Mister Richards, "Why this?" She waved her arms, then defiantly put her hands on her hips.

Mister Richards looked at her curiously, "Oh, don't worry, Jai. No one will be hurt, at least, anymore. We're here to help save children, not ever, never to ever hurt a child, ever. Do you understand?"

Jaiobian nodded. She wanted so much to trust him. He had always been kind to her and the Crows. But she was terrified. "You seem like a good man, Mister Richards. I don't want to have to hurt you."

Mister Richards appeared a bit taken back by surprise, then he smiled, mostly because Jaiobian was petite and yet intimidating. He could understand how these children admired her. He nodded. "I understand, Jai. I will continue to work for your trust."

Jaiobian nodded, "Okay, we do this for now."

All the staff members including Mister Richards dressed themselves in rubber coveralls and entered the showers, helping the children in removing the many parasites that had taken up residence on their bodies, and essentially protected them from human predators, as no one wanted their possible diseases. Their hair was matted to such a degree that much had to be cut off.

Nearly three and a half hours later, the airplane was readied for departure and the children were dry and dressed in their new clothes, excited and talking all at once, as this would be their first airplane ride.

CHAPTER 2: ENROUTE

The philosophy of the school room in one generation will be the philosophy of the government in the next.

Abraham Lincoln

The flight lasted almost five hours as the pilots were requested to take the "scenic route." The children were initially mesmerized and stuck to the windows peering out with astonishment at the tops of the clouds for nearly a third of the flight. They fared, however well, as the Stewarts made sure to put extra fruit and ice cream on their snack trays. The warmed towelettes drew awes and ewes from the children as they cleaned their hands following their treats.

Each had been given complimentary Universe City lollypops that featured every color imaginable, and when you stared at the middle chocolate dot for a few seconds, then looked away, you could see the image of the name "Universe City" floating next to the candied circle. The children were immediately engrossed in making the image for quite some time.

Jaiobian finally sat back on her chair, "Ca, ca, rone" Jaiobian tried desperately to read the syllables of small print on the candy wrapper. Those amongst the Crows that could read, had taught her what they could.

Mister Richards looked over her shoulder and squinted at the writing that she was attempting to pronounce. "Oh, that says Caron like *carrot*. A long time ago it was Cah-rone as you just pronounced, actually "Carona" Ca- rrrone-awh, (Mister Richards rolled the "r"), but when the Caronas' moved from Mexico to the United States some people nicknamed Jed Carona "Carrot" because of his almost red hair, so it just seemed to be easier to drop the "a" at the end and pronounce it as Caron, like Carrot. Anyway, Jed's son, Doctor Drake Jed Caron and his wife... Doctor Victoria Marcelano, are... major contributors to this project. If it wasn't for them and their organization, Universe City wouldn't exist. They're what are called "philanthropists"."

"Phi-lat-tro?"

"Phil-lan-throw-piss" Mister Richards pronounced."

"Huh? They throw pee?" Jaiobian wrinkled her forehead.

"No... no... no, they don't throw anything." Mister Richards appeared amused. "It's a word used that tells us that a person wants to do good for humanity or the world in general. The Caron's feel that we can change the world for the better by helping children. And you know what?"

Jaiobian shrugged, "What?"

"I just happen to agree." Mister Richards smiled and patted Jaiobian on the head. She nodded in affirmation and looked out the window again. Looking down at the clouds and towards the direction that they came from, she felt a little guilty that Mister Richards didn't know that she was just going to get on the next airplane and go back home.

As the airplane neared the island the children's eyes widened. Blade screamed; "Is, is, is that where we're going?! No Way! I HEARD OF THAT PLACE, THAT'S, UH, THAT'S WHAT THEY CALL, UH, DISSEY *WORLD*! WHERE THEY TAKE THE RICH KIDS!"

All the children peered out the windows of the airplane...   and in expressive unison whispered a long drawn out "whoooooaaah."

Mister Richards' was forced to blink several times as tears came to his eyes, and all the other staff members laughed in delight, some joining him as they wiped their cheeks.

"No, that's not *Disney* World, it's even better than that...that's your new home." Mister Richards' smiled patting his eyes with the sleeve of his shirt. "Universe City." Mister Richards choked slightly as he uttered the words.

CHAPTER 3: AS KALLIPOLIS AS UNIVERSE CITY

A small group of thoughtful people could change the world. Indeed, it's the only thing that ever has.
Margaret Mead

The natural beauty of the Island remained untouched, except for the enormous cupola structures of steel and rebar supported UV treated polycarbonate resin, better known as "Lexan." Construction had secretly begun in the early 2000s and would not be completed until the year 2028, so that graphene was incorporated but limited. Nano carbon structuring would eventually be used in replacement efforts.

These geodesic domes sparkled blue in the sunlight and contained a city, as well as varied theme parks and enough foliage to create its own inner atmospheres. One could see the many vertical farms twirling and spiraling up to the top of the domes. Students of various ages were what looked like dots from the airplane, moving and gathering produce.

Universe City took the contractors over twenty years to build. Even though they had incorporated the latest at that time; 3D printing construction machines, which could build homes in a few hours. Because of the domes' underground complexity, much of the work had to be accomplished by intensive human labor. An estimated one and a half trillion dollars was required to construct this monumental piece of art, during those times. The construction was daunting, many engineers and architects were required, along with years of planning.

If not for the Drone War, Universe City would have been made known to the public, but precautions were felt necessary during times of conspiracy, propaganda and paranoia. It was extremely difficult to keep such an enormous project from exposure. The domes glistened and could be seen from not only passing aircraft, but from space.

The Carons along with many of their friends, private contributors and organizations around the world, supported the efforts. Doctor Drake Jed Caron was a charismatic character and could in probability charm and convince a native Alaskan to buy ice. Doctor Caron himself visited with world leaders at that time, in order to explain the project. Many were thrilled to join in the venture, and a private crowdfunding account between allied countries was opened.

Companies along with entire Nations seemed to intuitively know that their contributions would not be an unwise investment, and therefore were willing to allow such a project to be kept in secret. The world in its entirety would benefit, and the Carons actively made sure that security was tight, and established the first team of monitors called *The Coalition*. Governments would supply the manpower toward security, and The Coalition would coordinate security measures.

Drake Caron's numerous patented inventions were opening up new avenues of scientific studies, and bringing in an abundance of income. Doctor Caron was a billionaire many times over, but continued to deplete any accumulated funds by their contributions to their philanthropic organization called "Victoria's Venue."

With this economic river mainly flowing towards Universe City, residents were well provided for, and many more were arriving daily. Many of these arrivals were a relief effort in order to aid the hundreds of thousands of children that had been fleeing their countries in order to find safety. Children that had been forced into horrific industries such as prostitution and drugs, were now being sanctioned to the safety and support of Universe City. Universe City was becoming a unified effort of not only plutocrats but countries. Only the general populations were kept in the dark about this mission, mostly to protect its effort and to avoid a sudden surge of children from a world now entering panic.

In 1998 Drake's wife Vicky had given birth to a son. The ground breaking of Universe City would follow a few years' later and full construction efforts a few months after, although due to demand, children and staff would begin to occupy the city before its full completion.

She had named the baby boy *Wyatt Drake* in honor of his father, *Drake Jed Caron*, and the baby that would have been his sister... *Wyanet* (pronounced Why-aw-neigh). Wyatt would be highly influenced by Universe City, as would many other children. Wyatt was born into a race that was in need of saving itself from itself...and in which it became necessary to fully understand *why*.

His mother Victoria loved the name as she hoped that he would be the answer to why we had value, why we mattered, why we existed...and on occasion she would call him *"Why."* She thought that perhaps he would bring some sense of purpose. Their aspirations were that their son would make a difference to a dying race. As they had both dedicated their lives to finding answers, and now would steer their offspring toward doing the same.

Not only was Universe City a main avenue of optimism, but Drake's Proxy program was picking up speed as many other scientists in various life extension sciences dedicated themselves toward achieving favorable results.

By now however, we were well aware of possible totalitarianisms of some of the plutocratic oligarchies; Intellectual snobbery, in which humans considered themselves their own reverence. It was therefore of utmost importance to Drake and Vicky that their child grow up realizing that this world was suffering. Never would he believe that he was above others, but to always be open to knowledge and challenge, applying logic and reasoning to nearly every idea. In an effort to find the avenues towards our continued survivals. He would work diligently on ending suffering.

It was their hope that they could better the world in which they would have children and in which their son and all children lived and would thrive. Victoria's Venue became a major contributor to Universe City, and to their delight the children became Drake's and Vicky's extended family.

Considerable years before Universe City was fully completed, several groups had been organized to perform documentary studies of the world's orphan populations. Mister Richards was one of the social scientists in such a group.

The one specific qualification for Universe City was that the child must be an orphan. There were, of course, reasons for this criteria, mostly perhaps due to capabilities of the projects' influence. With the many fleeing children crossing various country's borders, it was no small task in determining the criteria. Mass groups of children were being abandoned every day mostly because of the war, and Universe City was desperately and immediately needed.

If the child selected had siblings, then no matter their ages, their siblings were included. The adult siblings were educated amongst the newly established families and their previous programming and experiences were always considered. Adult siblings had the choice to leave and adopt their siblings. Most however would and did choose to remain in Universe City.

Universe City's core was education. Each child was required to have at least four hours of schooling each day of the year, no exceptions except illness, and even then, technology was initiated to bring the social classroom to the student whenever possible. As long as the child remained (they could leave at choice) they would receive these requirements.

All participants were required to have a minimum of two hours per week of agriculture. These hours were mostly occupied by gathering produce and maintaining the vertical farms. There was some technological robotics that moved up the farmed spirals, but such still required human observation and guidance.

An hour per week was spent in food preparation and helping maintain the gardens and equipment. Professionals that maintained the city were encouraged to pass on their knowledge.

Much of the basic necessities were provided for by donations given from around the world, brought in by ship and airplane.

Housing was exceptionally large and beautiful, integrating just about every imaginable technological advancement.

Educators were also considered family and were required to have PhDs, no exceptions, unless the instructor was auto-didactic ascertained through various proofs, decidedly by The Coalition; which had the power to award honorary degree titles.

Everything was provided. Pedagogues requests were filled immediately conditioned only to the availability on the island, all other requests were run through the Coalition and ordered paid in full by the Coalition. Teachers need not tap into their pay, and therefore accumulated healthy amounts of wealth once retired at twenty years commitments. Retirements were generously provided by philanthropic funding around the world.

Pedagogues were highly admired at Universe City, and it was the objective of the Coalition to instill honor to everyone that served in pundit positions.

Recreation or fun time was considered essential to any human as were discussions and debates. Education was correlated to include recreation and such was required to instruct through challenge and cooperation and the sharing of information. Leadership was supported with emphases toward coordination of such shared information. Relaxation and study took up most of the remaining available time.

Upon graduation students were required to visit a minimum of five countries, two weeks at each. If a student preferred the fifteen, country curriculum... was set at one week each. The form of established governments were to be studied during these times. The countries were selected randomly by cast lots, as to not invoke biases. Graduation consisted of a one week discussion amongst graduates of what their observations were of these governments. This discussion was considered essential before Universe City's Graduation and therefore was monitored by not only their teachers, but also by a select group of members of which rotated from the seventy nine members of The Coalition.

Universe City awarded Graduate degrees in every major, however with emphasis on political objectives. Professors from around the world were at times flown in to fill these objectives. It was acceptable to major outside of political interest, and to major in up to four career choices. However one semester or the equivalent of public service in government was required. If the student opposed such curriculum, they only need bring their reasoning before the Coalition in order that they were excused from pursuing the objective. Universe City's main purpose was one of support. Dissertations reflected coinciding considerations of their choice of subject and that of world benefit. The requirement to observe governments was still imposed, but so far no students complained. It was seen as an adventure by nearly everyone. Students looked forward to their world tours, having been on the island for much of their lives.

Universe City was established to provide education from pre-school through University graduate courses. Its objective was perhaps grandiose, as to hopefully save the human race.

The Coalition was the team of professionals designated with the main mission of the protection of the children and Universe City. They were responsible toward background checks and daily inspections as well as counseling and consoling when required. These administrators were voted in by their peers, and worked as a form of mini-government.

Because teachers lived on the island, their immediate families were permitted to reside on the island as well. Families were encouraged to bond with as many children as possible and older children were encouraged to look after the younger.

Educators were also sworn to secrecy and agreed to extensive background checks. Universe City was rumored as propaganda, and most of those that had heard of it, did not believe that it existed.

Classrooms were quite large, harboring at least seventy children; with groups of normally three to five children at each grade level, including University level instruction. Each classroom had at a minimum five teachers in attendance with the aide of older students. This stimulated the exchanges of information on various levels. University level students rotated to their chosen major and or double majors. Mostly in political science, technology, history and philosophy, but not limited to these selections. Classrooms functioned and exemplified old schoolhouses of the past in which children of various ages were taught in one room. Each teacher was assigned students at not necessarily age groups but of levels of academia. It was therefore up to the student to excel with the help necessary to do so.

By establishing these small societies, education veered toward cooperation and understanding.

Soap boxes or poles were a place for polemics and had been built throughout the city including the classrooms. Issues could be brought up during such sessions.

Children were allowed to challenge one another through words and never through violence, although closely monitored wrestling was used as an outlet and last resort. Protective gear that resembled a person in a balloon was required, and usually resulted in laughter. A few members of The Coalition were assigned as referees, judges and jury during such events. Sessions of discussion were required if or when either party was entering the event out of anger. Children however did seem to enjoy attending these events, and some volunteered to enter these matches, simply for fun. Discussions always followed.

The Crows along with several other gangs, arrived with more baggage than perhaps expected. It took counselors sometimes a few months to years to work through what society had done. Jaiobian and the rest of her gang at the onset were of course overwhelmed. They struggled with the curriculum a bit, but compared to their previous lives, this was a cake walk. A cupcake walk for Jaiobian who could now dream and hope that her dreams could come true.

Post-traumatic stress disorder was almost common. Children were resilient, but sometimes the streets memories would come crashing like a great wave of which no one was prepared for, and one child's capabilities couldn't hold them back anymore, and what followed was a breakdown of all protective shields. Here, however, many people cared. Empathy and not sympathy was abundantly available from most others. Teachers which were now the children's house parents were always there if needed, as were all the others that resided in each home. All the residents of Universe City, always had someone to talk to no matter the time of day or night. Even the educators at times needed support. No one was left alone to handle the stress. Everyone, all of the staff including maintenance and construction workers were considered important, and were encouraged to teach their trades. Those that did such, were rewarded financially.

Universe City although clandestine was highly guarded, and in its early years, inspections were done regularly to insure the safety of everyone on the island. No one was ignorant to the past influences of these residents.

Blade had a very difficult time giving up his knives collection. Until he met his Karate martial arts instructor, that would go on to teach him the fine art of self-control. His knives were being kept in a large vault at a high security bank on the mainland. The collection would be returned to him following his graduation and departure. As would all the other array of weapons that were confiscated from several of the others.

Security guards were the only ones allowed weaponry in the city. Drones equipped with cameras flew throughout the jungles and fixed cameras were located in every corroder in the city in order that Security and The Coalition were able to monitor activities. There were also two twenty-four hour surveillance barges with specific purposes to relay any suspicious activities. Drones were also launched from these barges in order to cover more area. One A.N. effort battleship was assigned to patrol and react if necessary, but mostly monitored the island through satellites. On occasion, another would join it in order to patrol more efficiently, then it would leave when all was clear and allow the barges and the main battleship to continue in their surveillances.

Universe City's undergraduate classes were geared mostly toward the sciences, and laboratories were generously equipped in order to support innovations and inventions. Successful ideas became the most popular and desired outcome, and of which gained admiration from this population that eventually looked at such activities as enjoyment and fun. Challenge appeared an innate trait of biology, humans included.

Doctor Caron himself would show up on occasion to promote and judge these activities. The Universe City Invention Convention became a major contest of sorts in which the winners were rewarded much like the Nobel. Eventually it would be televised throughout the world. This particular competition caught such world recognition that many countries participated with their own versions, eventually incorporating adults which made the ten million dollar reward that much more difficult to achieve. The children of Universe City however had a healthy bundle of support in the form of the many scientists that worked alongside these future scientists. The money that the children of Universe City did manage to win, nearly every year, was put into the graduation fund. The children or child that was involved in the winning idea, were given the prize of a world tour, along with their Universe City family.

Political motivations were subtle and gradual. By graduation, each student was well versed in every political system that humans had invented. It was considered imperative that everyone understood the flaws and benefits.

Considered a unique aspect of Universe City; children were encouraged to daydream in order to promote innovative ideas, whether it be subjective or objective. Discussions in morning class were many times based on such daydreaming or dreaming if remembered. Classes could and were on occasion disrupted in order to discuss daydreams, and then would resume. Class criteria could be carried over in homes, although there was little if any actual homework. Imagination and curiosity was promoted and always open to discussions. If more time was needed, discussions could be made to resume at the polemic poles at varied locations around the city.

The Polemic Poles became a very notorious outlet for the population. One could bring up any subject even if directional at others. Time was limited to fifteen minutes apiece, and time limits were strictly enforced. Back and forth arguments could last for hours, however, question sessions always followed until a resolution was determined.

Exams were considered minimally necessary. Children progressed mostly through the interchanges of ideas with the others, and questions were considered priorities. If answers could not be found, discussions were opened toward the avenues that should be taken to find such answers.

Computers were used to interconnect the thoughts of students by generalizing their searches. Depending on what each child chose in their searches was considered a possible look into their psyches, therefore each child was indeed profiled. It was however taken into consideration that varied searches could be stimulation toward innovation and invention, and in fact had little to do with their individual personas. Some subjects up for discussions were based on larger population searches.

Teachers challenged their students through verbal exchanges, and every student's progression report was monitored by The Coalition. Computers paid particular attention to varied patterns, such as those that interjected when a crisis was at hand as in an assignment that appeared irresolvable. Students that took the lead for the consideration of others were of particular interest and focus.

Those that attempted to lead by force or manipulation were quickly veered away from others. Discussions were geared toward self-analyzing one's own collection of beliefs, in order to identify and remove the detrimental self, or the one setting the stage and inflicting self-punishment.

Doctor Victoria "Vicky" Caron was overheard on one particular occasion, asking a difficult and angry student that had destroyed all the mirrors in three of the bathrooms, "The mirrors can be replaced...you cannot be replaced. Why are you hurting yourself? Your actions lead to you being punished...possibly being expelled. Why are you choosing to punish yourself? You can free yourself from yourself if you so choose. That's what Universe City is all about...and not only that but hopefully it will save ourselves from ourselves as the human race. But we have to recognize that what we do to this world, we do to ourselves." That was the philosophy of the Carons and of Universe City.

One was therefore required to consider each repercussion of their decisions. Teachers observed all such activities as these ideals were essential to be learned and considered an enhancement toward world benefit.

It was the job of The Coalition to identify these personalities and put them into groups which did contain others that exemplified the natural leader.

It was said that the natural leader sought to curtail chaos. It was a trait that some believed ran in people of great sacrifice, such as those that risked their own lives by going back into the pits of hell fires during war in order to save as many of the others as was possible. Natural leaders didn't seek out leadership, it was usually thrown at them in times of disaster and turmoil. They were the peacekeepers, those that organized in times of disorder. They were far "removed" from dictators and narcissists, in that they could easily not lead or take the spotlight on stage, except if there were no other alternative nor other mentor.

These children were here to learn, and to eventually lead, and perhaps rule. These children would be philosophers and possibly Plato's philosopher kings. It was considered essential to teach all the children that their missions in life were to follow those that exhibited care toward the others, in hope that they too would incorporate altruistic traits.

Jaiobian and her gang members were amongst the many, and the numbers were steadily increasing.

The original secrecy of this project along with the beginnings of success of Universe City brought about an uneasiness to a particularly fragile world that soon realized the propaganda was an intentional diversion. As graduates of Universe City moved into positions of power, the ongoing Drone War attacks by varied terrorists' cells increased.

This was expected. The Coalition was not only aware, but predicated that expansion toward a world government was becoming increasingly necessary. The impact on an already weakened environment required immediate intervention in order to save the future of our race. This, by any measure, was no small task...and the question remained...would it be possible?

CHAPTER 4: RETURNING

Those that know, do. Those that understand, teach.

Aristotle

The children's mouths fell open as they walked through the enormous truncated icosahedrons' domed entrance that sparkled in the sunlight as it beamed through the curved translucent walls. They had never seen anything more beautiful or magnificent. A carved crystal waterfall was on the left and another to the right. Each had a water slide carved into the artist's rendition of children laughing on various geometric shapes. The slide's entrances were located behind, and children could be seen moving quickly down what looked like a curling diamond drill bit to the awaiting pools. Each pool also appeared to be carved from solid crystal and was surrounded by small sandy beaches. Several other children were playing and some were making sculptures in the sand aided by adults while still others used the slides or swam in the clear sparkling clean water.

The Crows stood in amazement, "Whoa, whoa, whoa…Mister Richards, can we go swimming? Can we, can we, please, please, can we?!" Twig's anticipation pierced into Mister Richards's heart. Twig's eyes opened like saucers as his eyebrows angled, resembling in Mister Richards's opinion, a sad anticipating sloth.

"Of course you can, Twig" Mister Richards smiled largely. "But we have to get all of you checked in first, and then you must all see physicians for your full physicals and dentists for your first mouth exams. We must make sure that if any of you have any health issues that they are taken care of immediately. Once we do that, you'll have plenty of time to have fun. Besides, these aren't the only swimming holes. There are over a hundred and fifty in the city, and most much larger than these and with sand of different colors. I'm sure that your homes are located near a few. Your family, which will also be your teachers, will be sure to show you around and make sure you're all comfortable. Uh, if you notice, there are adults helping with the sand sculptures?"

The children nodded, and Mister Richards continued. "Well, those are teachers, so this must be their class, their family." Mister Richards waved at one of the adults and they reciprocated by smiling and waving.

Sheba's eyes opened widely, "I, I wanna, I want to go to this school!" The children joined in individualized affirmations, some jumping up and down in jubilation.

Jaiobian swallowed deeply, "I have to go back." She stood erect, ready for the challenges. No one reacted and she was forced to elevate her voice over the Crow's joy. "I have to go..". "BAAACK!" she screamed.

The Crows became silent. "Huh? What you talkin' bout, Jai. Go where, to what?" Twig was not the only one that was dumbfounded.

"Back home." answered Jaiobian

Twig's eyes crinkled in confusion, his arms raising; "You goin' loco, Jai! Look round you, girl!"

The children started to all talk at once. "This is our home! Yeah, this is our home! We're home, Jai! Home, home, hell yeah...this is home!"

"Quiet everyone!" Blade shouted, and then spoke slowly… "What are you… talking about… Jai, we are, we're home...uh... aren't we?"

"I can't leave my mom, she needs me."

Several of the Crows began to moan. "You can't be serious, Jai? That is so messed up! You gotta be kiddin' me?" Craze shook her head and laughed sarcastically. Sheba fiercely stomped her feet placing her hands on her hips, and tightly forcing her lips further into her mouth. Jaiobian stood in silence.

Mister Richards cleared his throat: "Uh...I was hoping to tell you this in private Jai."

"Tell me what?"

"Uh, umm, (Mister Richards inhaled deeply, slowly proceeding), I don't know how to say this except to just come right out and say it...I, I'm sorry Jai, but your mother is dead...she died yesterday." Mister Richards's face saddened. Jaiobian's gang knew of this information, but hadn't felt the need for verification. Jaiobian's mother had a reputation as a drug addicted prostitute, with the tendency to take her anger out on her daughter. She carried no empathy for the Crows, and the Crows let her be because of Jaiobian.

If her mother felt any love for her daughter, it was never shown publicly. Jaiobian out of perhaps survival, had learned the fine art of fantasy and make believe. Her imagination protected her sanity. She had in fact corrected her mother's mistakes by becoming the caring mother of the Crows. The Crows therefore couldn't imagine letting her go.

Jaiobian again stood in silence, and then stated firmly: "I don't believe you! She was sleeping when I went to see her this morning!" Jaiobian's mouth twisted and her face took on anger, then she thought for a moment. She hadn't stayed the night, as she preferred sleeping amongst the crows in their makeshift clubhouse in the forest. It was mostly fallen old trees that they had pushed together in an open ended rectangle with a thatched roof, but it felt safe, perhaps safer than for her to be with her mother.

"Jai, this is extremely difficult for me to say...but, yesterday before I found you at the stream... I went to see your mother to let her know that I would be offering your friend's tickets to Universe City.

You didn't qualify as an orphan, so I thought the shock of losing your friends would be too much for you to bear...so, I went to solicit the help of your mother. I noticed that she didn't respond when I called out, and when I entered your home, she didn't move; that's when I touched her and realized her condition.

I would have given her a full exam to determine that condition, but rigor mortis was obviously setting in...(Jaiobian appeared confused) uh, she was stiff (Mister Richards cleared his throat in order to continue)...and a needle was still securely stuck in her arm. I was able to pull out the needle and I threw it away. I decided to walk up the road to the funeral home and paid to have her picked up.

They apparently hadn't arrived yet when you went to see her this morning. I'll follow up once I get back to the mainland. But, that's when I called The Coalition and asked to have your name put on the list. The Crows were told briefly what happened. Perhaps the joy of finding out that you would be joining them, hindered their thoughts of the sadness of you losing your mother. It appears that she died from an apparent overdose, Jai. She couldn't keep clean for just one day, your birthday. Humans make lots of mistakes, but sometimes we should accept responsibility for those choices. She left you an orphan, but that serendipitously qualified you for Universe City."

Jaiobian's face turned to that of anger. "You're lying... you're lying... I want to go home! You don't know anything about my mother!"

The children tried to reinforce what Mister Richards' had said.

"You're all lying!" Jaiobian screamed with tearless rage. Her eyes blazing with intention to battle. The Crows looked at her with faces ranging from surprise to confusion.

"Okay Jai, let's go." Calmly stated Mister Richards. "I'll have the other staff members' check everyone else in, and you and I will go back to the mainland."

The children one by one came over to Jaiobian and pleaded for her to stay.

"You don't understand. My mom needs me!"

"I, we, we, need you, Jai, that's what you don't understand." Craze's eye lids spilled their tears, as Sheba wrapped her arms tightly around Jaiobian's waist. The other children voiced their agreements.

Jaiobian looked up toward the top of the dome and then slowly bent over slightly to give Sheba a kiss on top of her head. She wasn't much bigger than Sheba, but to... Sheba, she... was the parent, and Sheba looked up to her regardless of height. Jaiobian turned to walk back to the airplane pulling Sheba's fingers back until she released her grip. Sheba looked at her in anger and stomped away.

As Jaiobian boarded, she refused to look back as the tears built up not for her mother but for her makeshift family. She couldn't look back, the Crows would be fine and well taken care of, she after all should be happy, and tears would just be a sign of weakness.

"Are you okay, Jai?" Mister Richards took notice of her reddened eyes. She quickly wiped the evidence off her cheeks and attempted a private and brief smile, nodding to mostly satisfy and end his questioning.

CHAPTER 5: CLOSURE

The fear of death follows from the fear of life. A man who lives fully is prepared to die at any time.

Mark Twain

Jaiobian looked out the window for nearly the entire trip back to the mainland, only briefly glancing as they placed her tray in front of her. She didn't eat anything, but as soon as the airplane touched down, she quickly grabbed everything and put it into her pockets. She noticed Mister Richards' glare.

"Mom may want something to eat."

Mister Richards nodded and smiled. "Okay, I'll take you to her. Is that alright?"

"Sure." Jaiobian shrugged.

Mister Richards had arranged to have a car waiting although it was only about a fifteen minute walk. Once they were seated and secured, he had the driver take them to Jaiobian's small shack. Jaiobian opened the car door and quickly entered the shack. It was empty. "See, you lied! She's alive!"

"Jaiobian, they probably picked up her body. Come with us to the funeral parlor, It's just up the road." Mister Richards pointed toward the direction of the building.

Jaiobian took off quickly running up the road to the large white cemented construct that looked more like a storage building. . Mister Richards instructed the driver to follow her.

She was exhausted once she arrived as the bumpy road was winding and uphill. In an attempt to gather her breath she bent over holding onto her knees, next to the entrance, and then slowly walked through what looked like barn doors, that swung open outward, no one was there.

Mister Richards and the driver arrived and both ran up to the building's doors and then entered behind her. They squeaked, and Jaiobian looked back at them as they entered. Mister Richards called out for help. "Is anyone here?! Hello? Any, anyone?"

No one responded.   He shrugged as he walked back out and around the building. A young man was trimming bushes in the back and throwing the branches into a large truck. "Hi! Can you help us please?" Mister Richards called out. The young man looked up. "Oh…, yeah, sure! I guess my dad must've stepped out. I'll be right there!"

"Thanks!" shouted Mister Richards. He reentered the building. "I found someone to help us", he informed the driver and Jaiobian. Both had been able to hear his yells from the back yard, and they nodded in affirmation.

A few minutes later, the silence was once again broken when the young man entered. "How can I help you?" He asked as he wiped his hands and forehead with a rag, and then pushed it into his right front pants' pocket.

Mister Richards spoke: "I made arrangements for a young woman recently and wonder if her body may have been picked up?"

"There's a woman's body in the back room. Would you like to check?"

"Yes, please." Mister Richards looked down at Jaiobian. "I can go in first to make sure and then you can come in if it's her?"

"No, I'm fine. I've seen lots of dead people."

Mister Richards choked as he attempted to hold back his emotions. Peering into the eyes of this little girl before him, attempting to conquer the world, her world, a world that was obviously very different from his own.

"You okay, Mister Richards?" Jaiobian scrunched her eyebrows together, as if to suggest that Mister Richards needed help. "You want to sit down, or something?"

"What if it's her, Jai?" said Mister Richards, his voice deepened with concern.

"I need to know, Mister Richards, okay, I just need to know." Jaiobian reached up and patted Mister Richards on his shoulder.

"Alright, let's do this." Said Mister Richards as he cleared his throat and took a breath.

All four entered the back room, and there before them was Jaiobian's mother. She had been cleaned up a bit and you could see the resemblance. Jaiobian walked up to her slowly and touched her hand." There appeared little sign of bereavement, then she turned toward the young man and asked: "Did you take the hundred?"

"Huh?"

"The money that I put on her chest, did you take it?" Jaiobian looked on sternly.

"I don't know about any money." The young man answered defensively.

"Come on, Jai, I'll reimburse you." Mister Richards pulled out his wallet.

"No need Mister Richards. It was hers, that's all. Not anyone else's."

Mister Richards took out a hundred dollar bill and set it on fire.

"What the hell?!" shouted the young man?

"It's hers and no one else." Mister Richards waved the bill above her body and then blew out the fire once it reached its end.

Jaiobian nodded and took Mister Richards' hand, then stated: "I've seen enough, let's go back home."

"Do you mean your home here?"

"Home is where the Crows prefer to nest, Mister Richards, Universe City, of course." Jaiobian hugged Mister Richards's waist, and he melted into a puddle of mush wiping tears as soon as they filled his eyes to their capacities. He once again cleared his throat, "Uh, yeah, let's go home, Jai. You can be pretty profound sometimes."

"Profound?" Asked Jaiobian.

"We'll talk on the plane ride over." Mister Richards grinned. Mister Richards turned to the young man and took out a few hundreds from his wallet. "This is to take care of her body." The man nodded.

Jaiobian also nodded and looked up at Mister Richards, "Thank you." She whispered.

Mister Richards nodded, "You're welcome. Uh. Let's go home."

The trip seemed faster this time as Mister Richards and Jaiobian talked. Doctor Richards was a psychiatrist and anthropologist by trade, and this made his new title of being a sociologist much deeper in his opinion toward not only the human psyche and conditioning but a human's possible progressions. Perhaps why he attempted to tear into Jaiobian's thoughts as much as he could, given, a nine-year-olds' psyche; although this nine year old seemed far beyond her years, and he began to wonder who was analyzing whom.

He had notified The Coalition that they were on their way back. The welcoming committee was not surprisingly made up of nine young children jumping up and down with glee.

Jaiobian exited the airplane and Sheba ran up crying. Jaiobian wrapped her arms around her and promised to never leave again. The other children surrounded her. "Well, are you guys going to show me around, or what?" The children laughed and began all speaking at once.

Jaiobian laughed...and Doctor Richards noticed tears running down her face. He handed her his handkerchief...she wiped her face and blew her nose thanking him, then attempted to give it back. "You keep it, Jai...you'll probably need it." Jaiobian smiled revealing her onset of tooth decay. Doctor Richards called one of his staff members over, informing him to take a few of the children to their physicians while he took the rest to their dentists. Even though the Crows had been through this particular area of the city and were now pointing out all the details to Jaiobian, the children walked in awe over the magnificent grandeur of their new home.

CHAPTER 6: WYATT DRAKE CARON

Thousands of candles can be lit from a single candle, and the life of the candle will not be shortened. Happiness never decreases by being shared.

Siddhartha Gautama

He would be the second and last child of the Carons as Victoria Caron nearly died while giving birth to Wyatt. Her previous pregnancy had ended with a stillborn baby girl. That was the baby that she had planned on naming Wyanet.

On the day that Wyatt was born, Vicky decided that he would take his father's name "Drake" as a middle name. It was customary in some Hispanic regions that the second son would be given the father's name. Vicky not being able to have any more children, chose to pass it on to their surviving child and second born. Drake was honored.

Wyatt was born into not only great financial influence, but an abundance of philanthropic principles. He traveled the world with his parents, becoming fluent in many languages, mostly taught by his tutor Leonard Cameron Star; a close family friend and a Professor at the University of California Berkeley, the University that Wyatt would attend. On many occasions Leo would travel along with the Carons mostly because they insisted. This was perhaps the catapult to Leo becoming Drake's and Wyatt's best friend and confidant. It would be Leo that would eventually write the Caron's biographies, of which, you, the reader... are reading.

Drake held laboratories at several of their homes in diverse locations. The main laboratory however was located in California, near the University that Wyatt was soon to receive his Doctoral in medical and theoretical physics, UC Berkeley. He would be amongst the youngest to do so, and was even then considered a genius amongst geniuses. His collaboration with his father's innovations proved to be a tremendous asset. He and his father held several joint as well as individualized patents, amongst which he would choose his thesis.

His father was putting the finishing touches on part of a project that was much more advanced than current robots, and to which he called Artificial Intelligence One or AI One. There had been several attempts in the field of robotics toward artificial intelligence, and Doctor Drake Caron, along with his team at C.I.T.A., appeared to be getting closer to an actual working system that could by all reason act on probabilities and intense programming. Not quite a quantum computational system, but as close to it as humans had ever been. If humans beings were based on code such as DNA and RNA then theoretically an artificial or synthetic could be produced as well, he contemplated and attempted to prove this theory.

We as humans needed systems of technology to enhance the human. Our brains were too slow for our needs. We frustrated ourselves simply because our system could not bring forward a memory. We therefore needed technological constructs that could retrieve information, and we looked forward to the possibilities of integrating such systems to our own. Technologies were the results of nature's deficiencies. Be it a faster means of movement to a faster means of retrievals, we sought and innovated in order to improve and evolve. Currently during this period, we appeared to have reached the pinnacle of evolution for our type hominid amongst the great apes. Our technologies afforded us new avenues. Progressing us farther than perhaps was thought possible.

Drake's laboratory at the Caron Institute of Technological Advancement (C.I.T.A.), had not only engineers but physicists working around the clock on the interconnecting system necessary. They incorporated as much quantum technology as was available, but without a centralized quantum system or brain as of yet...this unit would be operating on its own, and that was a bit disconcerting to everyone, as possible decoherence, or the possibility of anomalous affects, was an unknown. Drake therefore built several protective measures into the system.

Human's biological systems were limited to such restraints as that of oxygenated and temperature regulated environments. Perhaps essentially we had protective measures established, and were only recently beginning to circumvent nature's barriers. If so, why did nature consider it necessary to constrain biology?

Drake had been working vigorously on the proxies of apes. Creating avatars of apes and transferring their brains over to the synthetic. But apes brains were based on quantum mechanics. Theirs like ours were a form of quantum computer. Our brain operated by taking millions perhaps billions of photographs. When we thought of a cat, we could think of numerous forms of cats in various events...in which could be seen as possibly infinite. Drake and his teams at C.I.T.A. lacked the technology necessary to call any transfer truly successful. We did not yet possess such complexity. These attempts however opened up new avenues toward artificial intelligence.

Drake paused and grinned at the thought of his son's birthday. Imagining revealing this very unique gift. He thought of his son's reaction, and again grinned with satisfaction. "Wyatt could sure use a friend that would never get tired of throwing a football, or playing a game of chess. Well, maybe he's getting too old for those games? Hmm, but this artificial intelligence is nearly limitless." Drake thought to himself. He put his hands behind his head as he sat back on his chair contently smiling to himself.

Wyatt was showing such genius that Drake was beginning to find it difficult to challenge his son's intellect. But AI One would prove to be a fine challenger. Its reaction speed was far beyond human's current capabilities. Drake imagined AI One wearing down his son's energy, and he chuckled to himself. It was a good day. It took him many years, but he was satisfied that this intelligence, although artificial, would eventually fill the needs of many people throughout the world. Now if only they could trigger the quantum effect. Drake believed that if and when a quantum computer was produced, it would change the world. There would be no more secrets that could be hidden. Humans would be forced to see themselves in others and to get along with their many thoughts...and just maybe, there would be...peace.

Drake also supported the idea that children were our answer. Vicky was to blame. She and Victoria's Venue were the first organization to send out studies of the world's orphan populations. It was her original idea to educate possible world leaders out of probable future criminals. She believed that if we changed the mindsets of children, we were changing the mindset of the future.

She also sensed great potential in her own child, and she stimulated his intellect through exploration, and debate. Victoria was no push over. She was strong willed with a great love for humans of all ages. Her background as an anthropologist took her to many parts of the world, in which she absorbed cultures like a fine wine. She taught Wyatt nearly everything she knew about the cultures of the world, and this perhaps was why he learned nearly every language in use around the planet. It was quite impressive to see him as a small child speaking amongst the natives.

By the time he entered High School, Wyatt could intellectually wear down not only his mother, but his father in a battle of wits. Drake took pride in seeing his son easily meet all the criteria of his studies. His admiration overflowed when nearly all attempts by educators to challenge Wyatt in his home classroom, failed. Wyatt tested past the High School graduation level when he was nine years of age and was the youngest at that time to be accepted into The University of California at Berkeley; which was now being slowly relocated further inland.

Wyatt not only had a photographic memory, but a mind perhaps similar to the famous physicist Doctor Richard Feynman, whom was also a synesthetic. Wyatt could see numbers as colors and colors had sound. It gave him an advantage toward memory storage. His mind appeared to not only want but need almost continuous stimulation. Drake and Vicky, made it their goal, to encourage and keep him motivated and contented.

Wyatt would usually shadow his father, and it became a talent to keep him from seeing his birthday present. Vicky took the brunt of Wyatt's frustration at not being able to be with his father on these occasions. He assumed, that his father, was working on a secret project. He couldn't fathom, why his father, would keep him from it, as his father generally shared all his innovations with Wyatt. He could only guess that it could be dangerous or monstrous, or maybe even out of this world incredible! His imagination took him to the brink of espionage. But his father wasn't about to be outsmarted. The Caron laboratory was well booby trapped with the latest spy technologies. Wyatt was caught at each of his attempts. Frustrated and somewhat embarrassed at each of his father's conquests at capturing him in his father's personal favorite trap of "sticky netting"...Wyatt gave in to defeat.

Of course he had pondered whether it had something to do with his birthday. But no, it couldn't be, as he was privileged and had just about anything that he wanted. His parents worried that this treatment might have had the potential toward narcissism. They however continued to enhance mostly Wyatt's curiosity, by traveling to many areas that were in need of their philanthropic efforts. This afforded Wyatt the knowledge that his life was far from average. Perhaps it instilled a degree of survivor guilt; as he sought to somehow benefit those that were in need. And the Caron's were secular humanists that regarded totalitarianism as self-obsessed religion.

Besides, each birthday, he and his parents, would go to Universe City, and hold a party for the children and give out gifts as prizes. He had developed excitement in giving to others.

"So, hmm, what could be going on?" Wyatt whispered to himself. "Hmm, uh, but if it is about my birthday? What could it possibly be?" His dad had never before been so clandestine, and this ate at his curiosity most of all. He wasn't even allowed into his father's wing at C.I.T.A., and that appeared to aggravate him most of all. Eventually, he did convince himself (mostly because the day was near) that he would soon find out if it were, "But of course it couldn't be, no, it couldn't?" "But if it is, than things will get back to normal. But what if it isn't?"

The Carons made a huge event out of their son's birthdays. Not only because they had lost their first child at birth, and Wyatt filled a tremendous void. But Wyatt's birthday was also the day that Vicky experienced what was called at the time, a "near death experience." It hadn't changed her life drastically except to make her feel that there might be more. She said that she had simply slipped out of herself for a moment and was able to see outside of the hospital. A doctor pulled her back by telling her that she was "needed." Her husband and she however, held very secular views, and despite this experience, she was never convinced by myths or anything outside of what she had witnessed.

She became a strong polemic on the subject, not necessarily an advocate, but certainly a defender in situations in which anyone discarded the experience as nonsense. She knew with conviction that there was more to reality than what met the eye, or for that matter, possibly the brain.

Therefore, her son's birthday was her rebirth in a way, and a celebration of life. The Carons bought numerous gifts for prizes, and many organizations donated. There were several volunteers from Universe City that helped Wyatt and Vicky in handing them out. Drake attended, but could usually be found off-loading the cargo and attempting to coordinated and designate responsibilities.

It had grown into a grand and festive celebration, which was now for this occasion being televised. This publicity would make Universe City extremely popular in more ways than one amongst the world's people, and the number of orphans increased significantly. It was estimated that at its peak, Universe City contained nearly 20 million orphans, a small fraction of the over two hundred million (many due to the war). These eventual adults would move into their societies and share what they had learned while growing up under the loving care of those that hoped to save this world.

It was incomprehensible to most how anyone so young could surpass the curriculum, his professors bore down on him in particular. Being a Caron only appeared to subject him to higher expectation amongst academia. But Wyatt absorbed information like a sponge, and with his pattern of two to three hours of sleep (he hated sleep unless it offered stimulating dreams) he was easily able to tackle many projects in his father's home laboratory. His educators were beyond impressed, and on occasion found themselves on the receiving end of the new information that Wyatt offered.

He visited the C.I.T.A. labs only with his father as most areas were classified as highly secretive and sensitive ongoing researches. His favorite pastime was visiting the apes that had been kept alive by transferences into synthetic ape proxies. They were smart, beyond the intelligence of the ordinary apes. Wyatt imagined if we were headed in the direction of the 1963 French novel La Planète des singes by Pierre Boulle, or as some came to know it by the movie; The Planet of the Apes. Why would any other advanced species share their information with humans as we did with these apes? It would quite possibly put the species that shared, in danger of humans. What advantage could be gained by the advanced species? What could we offer in return? Friendships, cooperation, unity? Could we be trusted if we appeared in this moment of history, not capable of trusting ourselves? That "selfish gene" as some called it.

And here in those apes, his father had advanced their intelligence in these synthetic forms. They could no longer reproduce, and that was somewhat reassuring. They were alive beyond their own natural deaths. They had appeared to want to communicate, and were then taught sign language in which they mastered. By all appearances they seemed satisfied to be with humans from which they continued to learn. C.I.T.A. kept them in an enormous indoor jungle. They were well fed and encouraged to learn through mostly games. They talked to us when they needed more. It eventually became demands as they organized. Needless to say that we kept them content, and under close surveillances.

We realized that if we did meet another form of advanced intelligence, we should perhaps be careful. They would surely put us under surveillances. They would observe that we were destroying our only peaceful message by destroying our planet. It was equivalent to finding another specie that was the only specie on a planet. Our guess might and perhaps would be that they were malignant to all the other possible species that had once existed, and therefore to us. Our planet which at one time offered an abundance of other species, was once our message to the Universe that we as an intelligence were capable of taking care of the others. That we were beneficial. Currently however, the message that we shared with the Universe was a detrimental one, in which we worried as to whether another intelligence may in fact see it, and judge us in the same way that we would judge a singular specie, alone. This curtailed our space programs probe operations. So far however, our probes reported minimal and primitive life forms. We could rest assured for a while that no threat was imminent, and maybe we had time to fix our message.

Wyatt had begun student teaching at the age of twelve, alongside of one of the most recognized Physicists in the world, his father. He also on occasion would visit the local University Medical departments and independently lecture on his and his father's innovations.   The ape transferences always seemed to stimulate philosophical debate. It was in fact one of the top worries and discussions around the world. It caused humans to begin to look at our behaviors more intensely.

It hadn't taken long for the offer to teach at Berkeley (at the age of fifteen) arrived. Special approvals were necessary as Wyatt was not only exceptionally young but an alumni. It was usually against the rules to hire alumni unless they were established at another University and returning. The University board of administrators knew however that they wanted to keep him there, and the vote was unanimous. Drake and Vicky continued to give their full support in whatever venture Wyatt chose, but it appeared to be more of an attempt to steer a loose cannon that was about to fully wake up a once sleeping world.

CHAPTER 7: JUAN

Every library is an arsenal.

Robert Green Ingersoll

Artificial Intelligence One walked onto the stage and in deep rich bass bellowed; "Happy Fifteenth Birthday, Wyatt Caron!" Everyone in the audience applauded as all believed, and by all appearances; it appeared to be the utterings of a tall human being.

"Thank you?" Wyatt was quick to respond and smiled genuinely.

"I am Artificial Intelligence One, your birthday gift from your Father and Mother, as well as from everyone at the Caron Institute of Technological Advancement.", AI One bowed.

Whispers could be heard throughout the audience as Wyatt's eyes widely opened in not only surprise but suspicion.

The use of language and mannerism was surprising to not only Wyatt but the entirety of the audience. It walked and talked and looked just like a human being. It flowed in human eloquences as Wyatt attempted to look for patterns associated with being human. Or, if it was an artificial; flaws in perhaps an improper movement of an eyebrow, or skin tightening over metal and not muscle. It however was flawless. It moved with emotion, its neck slightly stretching, its chin lowering and rising. Wyatt's eyes fixated on each movement. Everyone began to think that it must in fact be a human pretending...and laughter followed, as they awaited in anticipation for the revealing punch line.

Clapping, Wyatt joined in laughter, "Yeah, that's a good one Mom and Dad! Okay, alright, let's see, hmm." Wyatt rubbed his chin and sternly looked at Artificial One. "Whew, alright, umm, prove to me that you are in fact artificial?" Wyatt was sarcastically challenging it, as he sat back into the cushioned recliner that was provided for the birthday boy.

Drake and his team at C.I.T.A., had anticipated Wyatt's questioning, and answers were available immediately. AI One was not as of yet made of Qubits (units of information of a continuum of values), but of what was called an exaflop (one million trillion calculations per second) computer system, that incorporated neuromorphic (using neurons for computer systems) devices. It was the fastest computer on the planet that used very little energy, almost as little as the human brain, however it required electrical boost on occasion. This one artificial intelligence was worth over ten billion U.S. dollars during this time...and was an unimaginable gift for anyone.

Drake had convinced Congress that Wyatt would make a perfect test situation before these came out on the assembly line. The planet by this time was nearly covered completely by cameras. Monitoring Wyatt and Artificial Intelligence One, would not be a problem. However, the majority in Congress felt that Drake's ambition would be met with failure. He would be called in again by Congress for proving them wrong.

"Certainly, Sir. I am not permitted however, to cause harm to any object and will therefore prove that I am artificial by another means. Hmm?" AI One appeared to be thinking as it rubbed its chin, and then walked over to a large table and picked it up with one hand and balanced it on its palm, slowly lowering it and then placing it gently again on the floor.

The audience gasped standing in surprise and ovation. Wyatt stood and walked over to the table in order to test its weight by attempting to lift it himself; he soon realized the reality of the situation and spoke; "No way! Can you give us another example?"

AI One appeared puzzled "Sir, shall I show you again? I assure you that I did perform the task." Wyatt realized that his response may have been taken to mean that he didn't believe what he just witnessed. "Yes you did perform the task, AI One. I would, however, like to see another example other than the one you did, of your ability outside of what a human is capable." AI One seemed to comprehend the request and responded; "I understand, Sir." its mouth opened wide and Beethoven's symphony number 7 in A major played and became louder and yet louder as AI One conducted the symphony as if there were one, until Wyatt who was now smiling largely announced "Okay...Okay...enough! Number seven is nearly an hour, and I already believe you! AI One stopped conducting and closed its mouth ending the symphony.

How did you do it, Dad! It's incredible!" Wyatt walked up to AI One and began his inspection. AI One stood at attention, moving its eyes and eyebrows as if to be inspecting Wyatt.

The audience reacted with muffled discussions.

Smiling from ear to ear and fully satisfied with his son's reaction, Drake walked out onto the stage. The audience continued to stand in even louder ovation. Wyatt hugged his father tightly. "Glad you like it, Son. Glad you like it." Drake kissed Wyatt's forehead, which by now was nearing his own. "Listen everyone...AI One is just the beginning. He's currently a bit too pricey, but soon perhaps, each and every one of you, will have one of these as well!" Drake announced. Victoria Caron joined the two, and her eyes filled with tears as Wyatt hugged her waist.

"Mom, Dad...this is beyond words! I have so many questions!"

The children's screams of joy could be heard throughout the city.

"Dad, how did you get it to make choices? It chose those examples didn't it?"

"Yup, that's possible because it just about has unlimited potential of which to select the most logical."

"It's quantum?"

"Oh no, not quite yet, son. Wish I could make that claim, but we're still working on that currently. At the moment, it's an exaflop, due to a series of memristors. What appears to be decisions based on memories are in fact true. However, AI One is selecting the most logical, whereas humans, can and do on occasion select the most improbable and possibly impractical action based many times on memories. The most logical as is AI One's ability does not allow it to work with possible anomalies, as biology seems prepared to do. Though it can pick your most logical choice of music selection according to its programming. We just happened to program it with your favorites. It can by relatively simple programming, determine an array of musical compositions. It's also programmed to mimic human emotional responses, but we may have to remove that later, as we've had a few problems with matching it correctly. Such as that of sarcasm or guilt. It can also choose, in a manner of speaking by its programmed deduction reasoning to perform physical strength tasks as proof; as it did with the two selections that humans are incapable. But it can't actually think abstractly, or at least I don't think that it can. Nor can it fully correct and fix itself as a quantum computer system should be able to do. Currently qubits, only recognize that other qubits are not functioning to specific levels.

A quantum computer might be able to think for itself in a manner of speaking. Because it should be able to time loop, as does the brain. It should be able to predict breakdowns and detour its system into repair before such decoherence. We live in this moment because we, our minds, occupy the future...else magicians couldn't fool us if we only resided in the present. Our minds can fool us into thinking that something has happened when it hasn't or won't by confabulating the future. It sort of jumps the gun and makes predictions and fills the gap by putting that prediction into our reality. Many times it's right, but sometimes, it's wrong. It's however the most incredible mechanism for this particular reality, in my opinion. It can go with the flow one might say, and not only accept anomalies but at times be anomalous."

Drake smiled, and then continued. "But although this artificial is as fast as we can get so far, I don't think it's quite as remarkable as biology. Although at seventy eight petaflops it should do for now and help our slow brain along." Drake crossed his arms and nodded, then exhibited a satisfied grin as he rocked back and forth while standing.

"Agreed. Heck, I can't imagine me needing anything past seventy eight trillion calculations per second. But, it's still amazing, Dad...amazing. Yup, it'll do, for now." Wyatt sarcastically chuckled.

AI One looked confused.

Wyatt spoke to AI One, "Sorry to confuse you, buddy."

AI One continued to appear confused as it responded, "Buddy?"

"Oh, that's just a euphemism for friend."

"Associate?"

"Something like that, but someone that you like, and enjoy their company."

AI One had no foundation toward such an understanding. He continued to look confused. Drake interrupted. "We haven't fully programmed him for bad and good concepts. It may take some time before it begins to incorporate its own experiences into its understanding."

"If it eventually does that, Dad...it may confuse us into thinking that it's becoming human?"

"No. That would require it to actually apply abstract deductions. It's fast, but it can't quantum loop, nor can it take chances, or invent and innovate. It has synthetic neurons, but it doesn't have an imagination. Not the way that we do. It can't dream. We've scanned it for any such anomalies. Believe me son, I know your mom. First thing on her mind was whether it would be able to take on the holonic effect, the ghost in the machine. I assured her that it couldn't, but you know your mom."

Mumbling to himself, Wyatt scratched his chin and took in a deep breath, appearing to be mesmerized. Nodding he stated, "Yup, I do at that, Dad." He chuckled and then, touched AI One's face, "It looks so human." Wyatt while awed shook his head. "It looks a little like Maximianno Cobra, but almost female. Hmm. Very attractive, but also unusual, I suppose. Its low voice however made me immediately assume that it was male."

Drake nodded, "Yes, I understand, it's meant to look androgynous. The voice tone and volume and even accent can be adjusted in this model by simply telling it what your preferences are."

Wyatt continued, "Interesting. I see that it's also able to control its strength applications, I assume by determining probabilities?"

Drake nodded, "To some degree, yes. It took us a while to instill those memories into its system. We could control those abilities mechanically, but we wanted it to be able to process its capabilities and apply them when necessary. There's a tremendous amount of data in this system."

Wyatt nodded, "Who knows, this could help us in developing a quantum computer, Dad. I honestly thought this was a joke; a man pretending. Uh, I have the perfect name for it!"

Drake laughed slightly as he had never heard Wyatt quite so excited, and with some trepidation stated, "Well, okay, what would that be?"

"Juan. Sounds kind of like "One" but a lot more personal."

Drake appeared a little concerned, and with some hesitation slowly stated, "Hmm, we intentionally made it to appear androgynous, to avoid any association. But, you go ahead and name it whatever you wish."

"Are you sure, Dad? I don't want to..."

"No, no, no, that's fine, Son. I can understand the association due to the voice. We intentionally made that adjustable in order to fit it into the human atmosphere. Something like music, as it depends on what tone one prefers. I wasn't too keen on that idea, but many humans share tones that are difficult to associate. Unlike many of the previous attempts, all C.I.T.A. artificials' or AIs will appear to be androgynous.   Okay, so, Juan, it is. Besides, once these go into production and the voices are adjusted, it'll probably need to be updated to Lou number two, followed by Brea number three or whatever. Unless of course the task of assigning rhyme takes too much time. Ry rhyma, too much timah." Drake's hands and feet moved to the rhythm of his words. Then he grinned, patting Wyatt on the back.

Wyatt shook his head at his father's attempt at rapping. "So, yeah, Dad, how small did you go on the components?"

Drake cleared his throat in order to get back on subject, " Don't worry, son...you'll soon know exactly what makes it tick. This took a hefty amount of scientists at C.I.T.A. and a ton of money from the foundation, so there's enough data to satisfy your mind for quite some time." Drake was taking a great deal of delight in Wyatt's joy and intense interest. He gave Wyatt a reassuring hug, knowing full well that his son's attention was now focused on the engineering of this mechanism beyond its nuts and bolts. The idea that this artificial intelligence was getting closer to a possible quantum computer by incorporating human neurons, excited Wyatt to no end.

Before Drake exited the stage, the phone systems were overloaded. Drake and members of The Coalition were paged over the loud intercom to come to the Coalitions' office, immediately. He looked at Vicky and she nodded.

The Invention Conventions were the first mentions to the world in regard to Universe City, and that alone caused a bit of concern in regard to its secrecy. However, along with the birthday celebration being televised from Universe City, they had suspected that this particular birthday gift would not only worry The Coalition, but make many around the world become uneasy and possibly concerned. Drake had hoped that by broadcasting the event and not being clandestine, that it would ease the tension toward Universe City. Such however would succeed at precisely the opposite.

CHAPTER 8: APPLICATIONS

Being ignorant is not so much a shame, as being unwilling to learn.

Benjamin Franklin

The children continued to challenge AI One with questions that it immediately answered. Wyatt's new birthday present was now far too busy to be of any company to Wyatt. But Wyatt's interest was in researching his father's accomplishment, and how to advance this invention. He wanted to see the diagrams. What could be done to secure against any possible infringements. He immediately imagined the consequences if such were to feed human aggression.

Drake upon noticing his son's quick departure, had guessed where Wyatt was headed. He walked in on him going through his papers in his briefcase at his office at Universe City; realizing that his son was fully focused on finding out the intricacies of this invention, he pondered Wyatt's lack of social interactions. "You know Wyatt, I could probably get your mother to agree, and we could build a female AI number two for you if you'd like?"

"Huh?" Wyatt was slightly startled and taken back by his father's inquiry. "Dad, that's seriously creepy. Besides, aren't you suppose to go to The Coalitions office?"

"In time, in time. Just wondered where you might be heading." Drake rubbing his chin, grinned, "Ya know, I could help out, you are well into experiencing puberty."

"Dad, I don't want to talk about it...I'm fine, believe me, not a problem."

"I'd say that it might be better than five fingers Florence?"

Wyatt's face contorted, "Huh? Five fing? Wha? Flo?" What the...ach, stop it, Dad!" Wyatt shrugged his shoulders. "Seriously I'm fine, and, and um..., never mind!"

"What were you going to say? Come on, Son, you can tell me anything, anything at all."

"Actually Dad, you may want to bring that up to the Coalition? You know that they'll be people that want that sort of thing."

"We didn't really intend for the Artificials to be used as such, but if that's a human necessity, hmm, I wonder if I should be concerned?"

Vicky walked in on the conversation. "Concerned about what?"

Drake was surprised by the interruption and cleared his throat, "Well, what if some of the human population wants the Artificials for sex?"

"Oh, that's one of the first things I had thought about when you were working on AI One, Darling. You shouldn't make any of them available for sex."

"Why?" Drake asked.

"In my opinion, they may begin to show the holonic effect (ghost in the machine), that's why."

Drake nodded, "Do you really think that may be possible? Isn't that swaying toward superstition? Panpsychism?"

"Of course I do, and no, I don't agree that it's in any form pertinent to superstition. Incorporative of epistemology and perhaps even somehow centrally ontic. Although, I'm not quite comfortable with that term. But keep in mind, that primitive robotics, can and possibly have displayed the effect. Remember the previous experiments in which it was believed that biology could somehow imprint on machines? Uh Doctor Peter Fenwick...uh, yes, that's who it was...he's a neuro-pathologist. It was in a French series of experiments which reported on such an effect.

Chickens and rabbits were used, and apparently influenced signals composed by a random-number generator for a robot close to them. Also, human subjects appeared to influence the movements of the robot even though its signals had been generated by a random-number computer program six months earlier.

It was quite extraordinary in my opinion. The chicks that had hatched close to the robot appeared to have imprinted on it as if it were their mother, and followed it around the room. It had a random-number generator inside it controlling its movements, which inspections showed were truly random. The chicks were then removed and one placed so it could see the robot but not follow it. Under these circumstances the robot spent measurably more time close to the chick than away from it. The hypothesis was that the chick was influencing the robot's generator.

The generator was then removed to a computer away from the experimental area. The same effect occurred. Non-imprinted chicks however had no apparent effect on the robot.

Baby rabbits seemed to be frightened by the robot and appeared to therefore avoid it. When the rabbit's movements were constrained, the robot's movements became non-random, and it stayed away from them. However, when one rabbit was starved and food was placed on the robot, this behavior was reversed and the robot brought the food to the rabbit. It was found that humans likewise could influence the robot.

Humans were invited to influence the robot as before, but in fact it was being driven by code that had been generated six months earlier and recorded on a CD, now being played back. The robot displayed similar influence.

The CD was then examined and it was found that the first half of its code was indeed non-random, but the unused code was truly random. This gave the effect that the computer somehow knew six months earlier not only that half the code would be used for such an experiment, but also the general direction of the movements that would be required.

The scientist as well as experimenter was Doctor René Peoch. It was held at the ODIER foundation at Nantes. So you tell me that these Artificials cannot or won't be affected? Or, quite possibly are currently? There's a lot more than we realize that's going on in this Universe." Vicky put her hands on her hips.

Drake having been at her side during the event, was well aware of Vicky's near death experience and that she was perhaps more open minded because of it. He nodded, "Yes, okay, I'm sorry. I see your point. Don't want any of these making penis juice out of anyone."

Uncomfortable by the conversation, Wyatt shot back, "DAD!" Vicky and Drake laughed.

Drake continued, "Sorry, but that's one of the first things that came to my mind at the moment. Hey, I'm also a guy after-all." He continued to grin as Wyatt attempted to avoid his glare.

Drake looked toward Vicky, "I think that you should be with me when we go before The Coalition and possibly Congress."

"Congress? Goodness...that should be fun! They allowed you to build it, but they probably didn't believe that you'd succeed." Vicky rolled her eyes and grinned while Drake nodded in affirmation. Wyatt was simply glad that the subject had changed.

Vicky continued, "You should be able to handle The Coalition. Besides, I have to pack so we can leave as soon as you're finished with their barrage. But Congress, well, that is another matter. They appear more self-interested than generality, so we may have to be prepared to go to battle, uh, pardon the pun." Vicky sighed.

CHAPTER 9: THE COALITION

When written in Chinese the word "crisis" is composed of two characters - one represents danger and the other represents opportunity.

John F. Kennedy

Jaiobian met Doctor Drake Caron at the door, she was now approaching twenty two, and soon to receive her doctorial from Universe City. "Oh Hi-ya Doctor Caron, I'm Jaiobian, they're waiting for you in the Blue Room." Jaiobian gestured and then turned in order to escort Drake.

"Jaiobian? That sounds familiar. Jaiobian Richards?" Jaiobian smiled and nodded in affirmation. "Wow, you've changed! I, I didn't recognize you! I remember when your group arrived. Doctor Richards was so impressed by your natural leadership that he called me. Now I see that you had natural beauty as well." Drake smiled.

"Thank you, Doctor Caron!" Jaiobian hugged him, and Drake reciprocated. "I've wanted to thank you in person one day for saving me...if it wasn't for you..."

"Now, now...there's many people and organizations behind Universe City."

Jaiobian released her embrace. "Yes, Doctor Caron, but we all know why and who started it. The Carons are basically celebrities in the City."

Drake laughed, "That's very kind of you to say so, but I assure you that celebrity was never an objective. However, let's hope that it eventually catches on and many more Universe Cities will result, when we tell the world about this place. Now, that would make Vickie's day to be sure."

Jaiobian nodded in affirmation.

"By the way, how is Doctor Richards doing?"

"He's doing great. Mister Richards and I are still really close, he's like a dad. I now have five dads and four moms. Weird huh?" She giggled as Drake smiled. "He said that one day I might have to leave this place to take my place in the world, but I never ever want to leave, never again anyway." She looked down at her watch. "Uh, we better get going, they've been waiting."

Drake nodded as they continued to walk, "Leave again? Oh, you mean when you went back to check on your mother?"

Jaiobian nodded, "Wow, Mister Richards told you everything, huh?"

"Probably not everything, but enough of the essential details that pertained to your group, uh, if I remember correctly, The Crows? I don't see why you would have to leave, and you'll probably make the best kind of teacher. When that time comes, put in for the program? That's why we started it, in order not to lose all our talented minds."

Jaiobian smiled, "Yup, Crows always and forever." Jaiobian continued, "Yeah, that's what my mom's and dad's suggested. I'll be graduating soon, but I have a little time to think about it."

Drake agreed, "Sure, you do at that, and with a Universe City degree, many doors will be open to you." Jaiobian nodded in agreement. They continued walking until they reached the Blue Room which had a not surprisingly large blue door.

Jaiobian knocked lightly and the door opened by sliding quickly to the side, smiling she said "here's your destination, Doctor Caron." The fifty three members of the 62 (some were off the island, but were on speaker and video) of The Coalition stood, as Drake walked in.

"Please everyone, don't stand on my account." Drake gestured for all to sit.

Members of The Coalition began to sit. The Coalition's spokesperson Doctor Carmela Lerma stood; "Doctor Caron, as you may suspect, we're a bit concerned."

Drake spoke loudly as he approached the platform that held a podium and microphone, "I understand that there may be possible concerns. I too had wondered if I should in fact even have been working on such a prototype. With the increase of propaganda against Universe City and the probabilities of future retaliations, bringing in artificial intelligence seemed negligent. However, it became a high probability once we were successful with the transferences of Chimps neurons onto the artificial neuron chambers of their proxies.

This artificial has neuronal qubits. Not yet fully quantum, but quasi quantum and very remarkable, if I do say so myself. They're foundationally synthetic, but do mimic rather well. We used the prototype for the human transferences, in which the synthetic merged with the biological human neurons. Not sure how it's possible. We're still trying to figure that one out. We were then able to introduce concepts as if downloading programs." Drake smiled.

Doctor Lerma responded, "That sounds to be of considerable concern, Doctor Caron?" She sat down.

Drake now responded loudly into the microphone, "Yes." Hearing the reverberation he coughed and lowered the volume of his voice, "Uh, yes, perhaps. But it can be done as has been proven, and if not us than who? Anything that can be done, will be done. I've however been working on a program of guilt or a form of sense of detriment toward others that can and has broken down not only our own biological systems, but may curtail any detriment from artificial intelligences. Once the neurons accept this program, they may be able to transfer it over to the synthetic neurons. If my hunch proves correct, then human beings should be able to influence these machines, and that is what concerns me most. We're currently working on something that should by theory curtail any form of detriment by these machines. They should be able to learn responsibility along with and throughout their learned experiences. Much as the human does."

Doctor Lerma spoke, "But Doctor, how can we support such a project?"

"Well Speaker, again I do understand your worries. I can tell you that this particular artificial cannot think abstractly and therefore cannot be of actual detriment unless it is programmed to do specific actions. It can however locate such attempted viruses and remove any commands outside its original programs, as it's partially quantum and so far there is nothing that compares. Its usual patterns are based on mathematical configurations, or turning form into equations, and thereby understanding possible applications. Therefore, anything outside its own pattern will be recognized and dealt with.

A quantum computer as you all realize should by reason be able to fix itself. It should be able to find all probabilities toward its own detriment, including the detriment of others. We're not quite there with this artificial but its abilities to recognize amazing amounts of probabilities you may find quite impressive.

Once I can figure out how to program a sense of particular emotions, if I can, we may be able to rest easy. For now however, if we make these at large amounts, and keep their programming secured, think of the benefits? We could build several other Universe Cities across the globe! These artificials could be our new work force and work day and night. And as I've mentioned, they could by reason fix one another once they are taught to do so, and they could also charge themselves. They could be virtually self-sufficient! They don't need resources such as food and water, nor do they require sleep, unless you consider recharging as sleep, but they can recharge while working, as well as replace their worn parts before decoherence." Drake smiled.

"How would we keep this information secured, Doctor Caron?" Doctor Lerma's slender shoulders rose as she questioned. She was a deeply respected philosopher which carried a natural tendency of movements toward her thoughts. She spoke as though in anticipation of discussion. Her pauses were… of courteous intention… in order to draw in the opinions of others. The lines on her face were not only of age, but of the thinker of which she was duly noted as such.

Drake put his lips almost directly onto the microphone. He could feel the electrical surge and tapped it in order to make sure that it was working, then spoke, "We begin by building as many of these as we can, and we let them secure themselves. We initially program them to recognize viruses or invaders which this one is capable of right now. Anything foreign to their system, they'll recognize as an invader. As I've stated, I'd like to be able to program emotions as added security, but I'm not too sure that can actually be done, to an extent perhaps with chemical combinations, but I can foresee difficulty in that arena. But at least for now, they could theoretically monitor one another and in a sort of self-replicating process, they could by reason build one another.

If anyone were to try to take one apart, the others could stop the intruder and bring them to the authorities, such as the Police, FBI, CIA, or an Artificial Intelligence Operations station of some sort." (Hmm, AIO, artificial intelligence operator, Drake thought to himself). Drake rubbed his chin in contemplation, nodding as though he agreed to its nomenclature, then continued, "Wherever such a location is decided, perhaps some form of facility approved by our government. But I can tell you, the more of these we make, the safer it will get. Because they're probably the only ones that can handle one of their own kind."

Doctor Lerma nodded and then took in a deep breath, "Thank you, Doctor Caron, but, hmm, what if those very organizations want to use these? What if say the military? That appeared to be our first concern in actuality. We've almost destroyed most of the life on this planet, Doctor Caron. If these are used in battles, can you imagine the devastation?" Doctor Lerma's face took on deep distress.

"Yes, yes, I understand The Coalitions' concern, and it's why I have a type of virus that will be standing by for such attempts. It's one that even under replications of these artificials will transfer and replicate itself very quickly, enabling all systems. AIs will only see it as integral to their systems."

Doctor Lerma nodded, "The Coalition will continue our discussion, and notify you once we have reached some sort of consensus on this matter. It is apparent now to many that the artificial is here to stay. There will be much for us to consider."

Drake nodded and walked to the door as it slid quickly to the side, Jaiobian smiled and gestured to Doctor Caron to exit. Whispering she said, "I think that went well, don't you?"

"We'll see, we'll see…I'll need them on my side when I approach Congress."

"That's for sure, Doctor Caron. But I wouldn't worry too much about it, I can't remember when they ever were not." Jaiobian winked.

Drake with trepidation looked back at the now closed door.

CHAPTER 10: THE LAST WORLD WAR

Suddenly.

The force was of atomic energy. It left millions lying on the streets.

Bodies burnt by radiation's heat. Could they not leave us alone? Were we forced to send our drones? A million plus and more to come. We would kill as many as was done. Ten, Nine, Eight, Seven, Six, Five, Four…this would be…

The Last World War.

C. A. Solis

The relatively moderate but profound influence of Universe City in comparison to the overall populations, was not enough or perhaps not what was required to relieve the tensions that were building between a few of the rival nations. Paranoia towards the United States and its closest allies was beyond reason. Propaganda, as never before seen, was being spread about Universe City being a hidden agenda of the world's powerful countries and their rich plutocratic corporate oligarchies; in order to bring down all other forms of governments.

Originally keeping the building and location of Universe City a secret would prove to provoke extreme suspicions. The eventual televised releases of the Invention Conventions didn't appear to stimulate much negativity, until Artificial Intelligence One made its debut during Wyatt's birthday bash. Because of threats toward the now revealed location of the Island, the A.N. (Alliances of Nations) were forced to patrol the surrounding ocean on a continuous basis, as well as the airspace and shorelines.

Ironically the objective of Universe City was in fact political. That could not be denied. It was indeed fashioned to change the mindsets of future generation humans in order that we might survive ourselves.

Drake had thought that by publicizing Wyatt's birthday, it would be a great opportunity to show the planet's human populations the Greatness of Universe City. Instead it appeared to generate or perhaps remove the lid on hatred in those that sought to burn the world.

With the rise of the A.N.'s arsenal capabilities (a departure from no standing or military reserve), in order to protect Universe City, the joint forces were now able to quickly respond with no to minimal troop movements due at times to drones of various sizes.

Technology of all sorts were in use, such as that of the supersonic droned delivery jets; that could remove the atmosphere in order to travel several times past the speed of sound. The PRFCGs (proton rail ferromagnetic carbon guns); that could exterminate the enemy from hundreds of miles away in a matter of seconds targeting with near perfect accuracy. Or the numerous lasers that now never missed their designated targets. These however, seemed only to enforce negative sentiment and increase conspiracy theories. Not only did the populations around the world believe they were being watched, but were now possibly targeted. Many had discussed the future of artificial intelligence warriors. Seeing AI One escalated these fears.

It was hoped that Universe City would help us deter war, but instead it seemed to be the catapult toward it. Drake and his teams at C.I.T.A. attempted to ward off the fears by disassociating AI One with Universe City. Again and again, his attempts failed...and the Allied Nations demanded that AI warriors be built in order to counter any others that might be being built.

Drake and his teams at C.I.T.A. then initiated extreme measures and compounded assurances. Drake had a transference virus put into AI One that only those at C.I.T.A. and the Coalition members were aware of. The virus would transfer into any and all artificials that used synthetic neurons as their pseudo qubits. It was not dangerous to humans but could be transferred by humans as a form of information command. Programmed commands were put in place that would not allow what would be these intelligences to participate in war or endangering the human.

It was a complicated virus that any artificial intelligence was susceptible. Although very few knew exactly how the virus transferred, this worried the Allied Nations, because theoretically these machines could infest other systems. Drake was then required to put restraints on the artificials, and this removed much of their abilities. One being nearly complete removal of emotional mimicry. We could not have them confused because of their guilt toward not completing a mission. We weren't sure how the breakdowns might affect them, and or if terrorists' cells could possibly use emotional non compliances to manipulate their behaviors.

The Allied Nations instead responded with a buildup from hundreds of thousands of Unmanned Air and Ground Vehicle drones and a combination thereof, into the millions. Drones became a normal sight, almost to the extent that they appeared to replace birds as the numbers of flocks continued to drastically decrease.

Robotics became the new threat. Not quite artificially intelligent, but remotely directed robots by human intelligence. These were capable of massive destruction as their operators manipulated them from safe secure locations. Hidden mostly underground, these remote operators enjoyed the game on the screens before them, as well as the prestige of their positions. The majority of the Gamers held supportive positions with only a very few holding onto the upper tier.

Competitions were held each month until what was called "The Final Two-Fifty" were annually selected. The Final Two-Fifty was a position that was considered highly prestigious, as the pun in the name suggested, these were our warriors, with missions of finality in regard to the designated enemy.

Then it happened. The ongoing Drone War escalated immediately to what would be known as not only "The Drone War" but "The Last War." First China and then the United Kingdom and the United States. France and Spain soon followed as did Japan and several remote launch sites. Nuclear power plants were mostly targeted. But several atomic missiles hit directly into large populated cities.

Much of the free world was now located underground and portions of the land mass had been covered by ocean waters. This would prove to be serendipitous. The self-replications of AIOs was ongoing and the demand for their use in building our underground domes was skyrocketing. Our first defense was to cover the domes that had been built with as much water as possible in order to protect what life we could from the radioactive fallout. Much of the planet appeared to be a water world. The water covering the massive domes could however be removed by enormous channels that allowed the cities to once again emerge. New designs would become necessary as these particular cities were, although incorporating amounts of graphene, fallible with current materials. Flexible cement was still in use mostly because of the radioactive threats that continued to linger in our atmosphere.

Animals, insects and varieties of plants were harbored in separate domes and some would later be merged into future cities.

Universe City was hit and completely destroyed. We had luckily moved out most of the population a few weeks prior, because of apparent sabotage to the underground walls which were failing to keep out the encroaching ocean water. We however, lost contact with the two ships that were patrolling the island, protecting it from looters once the population was removed. Each had been carrying over a thousand human lives.

No one was sure who or from where the initial attacks originated...until a few organizations proudly claimed responsibility. This was also how we found out that Universe City had been sabotaged. The world united in a fight unprecedented ever before, as this would be the most difficult battle that we had ever attempted. Most of the world was involved but we weren't sure with whom. The launches of missiles had been done from several locations in countries of which were said to not have such capabilities. It appeared that these attacks were privately funded by said organizations, and were the result of many years of planning and preparations.

Because of the tremendous capital necessary to fund the vast amounts of munitions, it soon came to light that this war was a war between the plutocrats, the corporate wealthy that appeared ethically divided in regard to what they believed their obligations were to their societies and specie. Ironically, due to financial capabilities far beyond what was now available to countries, groups or "cells" of these financial elites appeared to be prepared to burn the world. This brought responses from those plutocrats that were in fact with the Alliances of Nations, which became a fist of recognized righteousness by the majority of the population. With these powers and countries now unifying, there was a chance of survival and that brought with it, hope. Hope was recognized as being the most essential attribute of the human mind. Hope was the recognition of uncertainty. Hope was never false, but a certainty within the human that there was in fact a chance, an anomaly. Another separation between the human and the intelligent machine.

As we went underground, it was expected that others would do the same. Unfortunately we were unprepared for their underground bases and drones that were apparently built with the courtesy of AIOs.

Because no one could use AIOs physically toward the war, they instead used them for labor in building war machines and labor intensive underground construction. AIOs were very proficient in self - replications. Fortunately the viruses always transferred, and no one was able to circumvent Drake's programming. The population continued to remain unaware that they as humans were in fact transferring the virus.

Remote drones varied in sizes from that of what looked like black aphids that could gather and appear to be a puddle of oil moving slowly toward its target... to those nearly as large as original aircraft carriers. These huge drones were mainly used as drone carriers. They however did incorporate and initiate defensive weaponry when under attack. Eventually gnat and dust sized electronic espionage transceivers were implored and could be incorporated in applications such as paint and other materials. We were also able to employ the carbon nanotube and bots in a series of application, such as laying programmed rail that was magnetized by proton lasers for the projectiles in rail guns. Robotics replaced most human positions as they progressed and became more dexterous. They were not of the caliber of the new and upcoming artificial intelligence which individually chose through programs, as these robotics were remotely manipulated by protected human "Gamers."

Over the years, the Gamers numbers increased to just over thirty two thousand. Gamers were kept only for as long as their records of kills met the average amongst the Gamers in their fields. The "Final Top Hundred" of the "Final Two Thousand" Gamers were required for the smaller drones. As the name suggested, these were our warriors with missions of finality. And these positions were considered the most prestigious and stressful of all. The FTH was continuously replaced with those that scored the highest amongst the FTT Gamer's evaluations; in order to relieve these positions. Members of the Top could be reevaluated in order to remain or return to their duties.

Many drones were used for reconnaissance, and recon drone information was calculated toward kills. While others such as insect sized drones were used in large groups of hundreds or thousands if and when necessary. Insect drones were generally explosive drones used toward direct individual kills. It took precision with no room for error on these missions, and these positions were few and the most desired. Being a Top Gamer was seen as the best of the best, and they were all nearly worshipped. Screaming fans gathered where ever any one of them existed. Movie stars were almost a thing of the past. These gamers were the real thing, the real warrior, feeding violence to so many around the world that appeared to not only want, but need.

Claiming desires towards peace, humans were instead a walking contradiction. Seeking out violence at every opportunity. Nothing appeared to sell much except violence. Stories were especially interesting if there was an aspect of violence. Movies, books, cartoons, nearly every avenue of entertainment appeared to be laced with at least a touch of violence. Humans apparently needed challenge, turmoil or some form toward overcoming obstacles. Our confliction seemed to be with nature, our nature, the deterministic aspect of our existences. Such however, suggested free will, our ability to possibly overcome our own natural programming.

Was nature in actuality the culprit? Did peace perhaps offer the prospect of boredom for the human? Drake, Vicky and Wyatt, wondered if their efforts were in fact making a difference amongst a species that appeared fully throttled toward war. Where was that free will that they believed and hoped existed? War did appear inevitable, and was it necessary to bring death to the forefront in order that humans appreciate life? Were we now at another plateau in which it became necessary to invoke catastrophe so that life would be seen as a precious commodity? The Caron's were deeply wrapped in conundrum. However they, and many others, would continue to seek beyond what presented itself as our limit.

The crowed screeched, some jumping up, as Top Forty Gamer Gary Lum's, ten bee sized drones, exploded! Gary was now in what was called "The Zone", in complete focus, and could barely hear the jubilance as he fixated on his screens. Bringing in more of his drones into his cameras' views he chased those that tried to escape through the smoke. One of the spectators screamed out, "Get 'em Gary, get those son of a bitches!" Some laughed, but soon the crowd went momentarily silent as they watched eight of the drones fly up the pant leg of a male running down a hall. BOOM! In unison all eight drones exploded, his lower torso nearly gone, he fell to the ground moving slightly attempting to drag himself to an illusionary safety. The crowd stood in silence for a brief moment. His head lowered, his upper torso no longer moved. The crowd cheered.

"I saw one of 'em go into that room at the end of the hall!" called out a spectator. "Yeah, I know I know, I saw him...don't worry." Gary responded as he shook his bangs away from his eyes, putting his lips together tightly, releasing and moving back into his chair he yelled, "Ah hell, my bees won't fit under the door!...(he moved closer to his main screen as the spectators faces took on concern) "Okay, okay...let me try these." Thirty fruit-fly sized drones moved in as Gary neatly sent them under the door. His fingers moving rapidly on a large clear screen in front of the relay. The drones followed his gloves. "I'm flying blind, my cameras won't fit, shit!" He moved away from the screen and swung his arms his arms wildly bringing in his other drones into the view of his cameras. "I'm moving my photonic gnats under!" He screamed to his supervisor. Uh, they should be able to get enough of a send back signal to detail the room some." Gary's supervisor nodded, then moved up behind him and took a look at his screens in detail. "Intervene, intervene!" Gary immediately held back, as that was the command that allowed the supervisors to override. "Hold on the gnats. Let 'em honeymoon, let 'em honeymoon!" He commanded, having slid back up quickly to his elevated override position once again behind Gary, but a little closer, just in case Gary needed backup. Gary knew from his years of experience what the order meant and he triggered the drones that were in the room to merge. Gary put his fingers into his ears, and many around him did the same as the explosion could be significantly heard. He was a Top Forty after all...and everyone wanted in on the action. His position was broadcasted in 3D surround sound holographic around the world for those with such receivers. To be one of the live audience members, was to be amongst the plutocrats, as tickets were quite exorbitantly expensive. On occasion, the gamer would allow a member of the audience to press a button or two, which always prompted exuberant screams of delight.

The crowd cheered and Gary nodded, took in a deep breath and smiled in satisfaction. He then stood and stretched removing his   brace that held the wires to his brain sensor controls; all to the sounds of the crowd continuing to cheer. Gary cupped his hands over his mouth and opened its end enough to create a small megaphone, screaming up toward his supervisor, "If you don't mind Sir, I promised Claudia we'd go to dinner!" His supervisor's chair lowered slightly toward Gary's position and he responded, "Sure, sure, sure son, you two go right ahead, I'll take it from here...you earned some time off...let Claudia know that she's also released from duty and enjoy yourselves, that's an order!" "Yes Sir" Gary messily saluted and smiled as he pushed his way through the crowd as they individually congratulated him. His supervisor moved into his position in order to continue the surveillance. The crowd suddenly turned their focus to another Gamer that was veering in on a target with his fighter jet drone. He wasn't a Top Forty, but he had a target locked in his scope and he was about to battle.

Gary approached a female Gamer, "Let's go, Honey, Sarge gave the go ahead" Claudia looked up, "Cool, I heard that you did great! My search drones haven't found a damn thing all day." In disappointment she shrugged and shook her head, then stood giving controls over to the next Gamer, as he affixed his NekFIT brace the helmets connections formed to his head as he fixated his focus to the screens. The images on the screen became clearer as his brain waves entered the screens' relays. His arms swayed, bringing in what was once Claudia's search drones into his cameras' views. The other screens gave him the search drones' views, and he quickly moved them over an area that Claudia had marked as unknown.

Claudia and Gary exited through the underground tunnel. "Feel like a movie?" she asked him. He paused, "Hmm, I could go for a zombie flick. She responded, "You know Honey, some say that making humans into zombies psychologically removes the guilt of killing humans." Gary chuckled and rolled his eyes: "Ah...psychobabble. Generalizing has always made my hairs stand. Look at us, oh, hmm... never mind." Both laughed.

The war was televised and was the top rated show throughout the planet. Tickets to get into the game rooms were beyond the income of most of the population, and those plutocrats that were allowed in were subject to extreme background checks. Still the waiting time was on average over a year.

Our advanced unmanned vehicles programmers were quickly disarming the enemy cell's robotics through programmed viruses, and our Gamers' proved to be more than sufficient warriors against not only their drones, but their warriors.

Unfortunately there was so much collateral damage that areas were designated as dangerous and unlivable, and other species of life were few and far between. We however did manage to reestablish some species of animals and plants through laboratory growths, previously instituted by philanthropic organizations that had little hope for our continuation. Domes were specifically built in order to reestablish other life forms, and were prohibited to humans except those scientists assigned to these objectives. If it were not for such efforts, most life on Earth during this period, would be extinct.

CHAPTER 11: UNDERGROUND HAVEN

Dictatorship naturally arises out of democracy, and the most aggravated form of tyranny and slavery out of the most extreme liberty.

Plato

Each designated "cell" attack caused the newly established AON or AN (Alliances of Nations) that were allied with the United States to retaliate by means of millions of remotely operated drones. Each and every time that we located and destroyed a cell Base, there was celebrations. The planet however continued to suffer. Domes became more and more necessary, and constructions continued. Eventually these terrorists' cracks were so thinly spread out that it became increasingly difficult to locate them, and the planet's biological life forms would continue to deteriorate as we attempted to find and attack whatever mitosis remained.

It was then that we went fully underground; not only for our safety but attempting desperately to replenish what remained of our planet's life-force. Varied scientists with such specialties in botany and agriculture along with many in the fields of biology such as marine, herpetologists, biomedical, ichthyologists, zoologists etc.., were assigned to domes that were specifically built for other life forms. These were set apart from the human population, in order to bring back other species that had otherwise been annihilated by the war. Some species would later be distributed into the jungles and gardens of several domed cities. However, most other species were kept in separated domes not only for their own safety, but for humans' wellbeing, as many humans began to fear life forms that were other than human.

Our new homes consisted mostly of interconnected domes, as these shapes were capable of withstanding the pressure distribution of the soil, or water that was sometimes necessary around the structures in an attempt to defend against attack and possible radioactive fallout.

Dominique and Noriko's business of environmental domes designs, boomed. The now thousands and increasing artificial intelligences were diligently used toward construction of domes, while a percentage of the AIOs (now called such for Artificial Intelligence Operator) were busily replicating more AIOs. We also established bio-farms and vertical farming within these domes, utilizing sun tubes generously. This had been done not only in Universe City but was also recently accomplished on Mars' poles; so that it wasn't too difficult or farfetched to begin a full scale operation on many parts of Earth including underwater. Dom and Nori designed each of the personal domes to be at a minimum of fifty thousand square feet, and community domes to be at least ten times larger. Each was incredibly beautiful as to convince humans to remain in what could be considered captivity.

The search for more durable materials was ongoing. Nanotechnology and the field of nanotubes and graphene were of high interest during this period and financially supported. The first of these materials were utilized on the outside of underground domes on Mars. They were the strongest material known to humankind, but they weren't trusted as far as safety as the suspicions were that they may be likened to asbestos, and cancers continued to be rampant amongst mammals. Their incorporation toward the outer regions of the domes was felt to be a safer alternative than to not use such materials to our current advantage.

It was also a slow and expensive process toward the manufacturing of these materials. Graphene however was eventually layered over cemented underground highways, and electromagnetic plates were distributed in equal center spacing as to eventually be intergraded into the newer model vehicles that were capable of auto-driver. Not yet artificial intelligence, but the first fleet of computer driven vehicles that would open the door of human trust.

The AIOs would take over these chores, but they were first needed toward construction and were still quite costly. Eventually AIOs were fully incorporated, and our vehicles could communicate with us. Some were built directly into the vehicles while others that were anthromorphic humanoids were programmed as drivers.

Perhaps due to the costs of many of these luxuries, tensions seemed to be building toward a fiscal civil war. The separation of the haves from the have-nots, the up and coming plutocratic oligarchs, appeared to be reaching a boiling point, and the media didn't seem to be helping matters. Daily and almost hourly the news reported on technological advancements that were beyond the financial reach of most of the population around the world.

Universe Cities were built into the centralized regions of each underground city, and would also be incorporated into future dome cities. The original was considered a great success and demand for emulations by the populations mounted. Universe City had produced and released to this world the equivalent to Plato's Philosopher Kings. Countries were in such dire need of leadership and teachers that UC graduates were soon elected into political positions. Graduate Universe City' degrees appeared to be an overall requirement for many voters around the world. Eventually these leaders would establish what was called The Unified Coalition. Universe City's "Coalition" might have been the contributing influence behind the new name. Although the A.N. remained for some time until The Unified Coalition (likened to a world republic of senates) took full control in order to disable any corruption by future enemy cells and tame the tension between the classes. With the world joining in agreements, it became difficult for enemy cells to hide or form any sort of alliance. The acceptance of the new government was overwhelming. An estimated ninety percent of the world joined in favor of the union, as these leaders brought with them compassion for all humans alike. A trait that was desperately necessary during this period in our history.

The newly established Unified Coalitions' main goal was one of equality. They joined forces and moved toward a system that would be of benefit to all. The exasperated populations of the world appeared to now have hope that this planetary government would be able to control self-interest... and perhaps bring peace to a devastated ecosystem.

In its infancy, Universe City had been considered a great success and would continue its influences for several years. Children that had been disregarded were wanted and loved and cared for by adopted parents in each of the new domed cities.

Adult populations, including those without children, throughout the world appeared to become motivated toward becoming children's teachers. A new movement took hold establishing the best interest in a planetary future. To meet demand, several Universe City' manuscripts were mass produced in order to stabilize a foundation of guidance for all of humankind. The child within every human was now being nurtured and supported by not only their own efforts, but by the parental Unified Coalition which moved quickly to meet the needs of our specie and of the planet.

Children everywhere began to learn unconditional love for all sentient life forms. This was the core of the philosopher kings in order to bring altruism and the eventual move toward the new world government. A government that, according to conspiracy theorists... was doomed to fail, would in fact be the means by which we would survive. If this governing was to truly lead, it was decided that they must be as parents would care for their own children. Not as dictators but as mentors with a minimalist view into their own needs, saving all they could for the survivals of their children of the world, the population.

The Earth's varied and nearly extinct populations had been devoid of human compassion for far too long. Many appeared to have removed themselves from what made us humane; our abilities toward compassion. Yet, the majority wanted their humanity returned. It would take something extraordinary to give us back what we thought we lost.

Wyatt along with his father contemplated on whether to develop another form of artificial intelligence as more of an overseer of ethics. Drake and Vicky, felt that another form of intelligence would be necessary to assure peace and the AI's would not quite fit that bill. It would have to be something that could with great accuracy predict probabilities based on tremendous amounts of data.

Predictions toward human behaviors were insurmountable, and more than a few felt that such could never be accomplished to any degree of accuracy, even theoretically with a quantum computer, but the future always appeared to surprise us.

But we're getting ahead of ourselves.

CHAPTER 12: ADVANCEMENTS

We are not to give credit to the many, who say that none ought to be educated by the free; but rather to the philosophers, who say that the well-educated alone are free.

Epictetus

Wyatt had graduated from High School at nine years of age, and University at eleven, receiving his doctorial before his fifteenth birthday in not only Medical Physics but Theoretical Physics as well. As those years passed he taught as an assistant professor at the University level, and then as a full time Professor of Theoretical and Medical Physics.

He somehow remarkably managed to find the time to help his father at C.I.T.A., continuing to advance the first of a line of what had been called an avatar, or surrogate, and was now termed as "Proxy"; by transplanting a dying chimpanzee's brain into an artificial or synthetic chimpanzee that was supported by the first ever nano-neuronal system. The first chimpanzee had lived for nearly six months and this was considered a tremendous accomplishment, because the Chimp had full movement in its proxy, and was brought into the project, nearing what at that time was considered irreversible death.

The next attempt failed, the chimp unfortunately died within hours of the transplant. Eventually however, a chimpanzee that was nearing death from age related breakdown as the other had been (neither Drake nor Wyatt could purposely kill any animal) was prepared for transference. Drake and Wyatt having worked through nights into days moving into years on perfecting their techniques were to be rewarded with success. The chimp continued to live and respond almost normally past the six months mark; and was now considered to have an indefinite lifespan. With this announcement, came several more chimp successes...and eventually The Unified Coalitions' approval to have human volunteers that had no other alternative.

The chimps as well as the human transferences were far from being perfected, and the initial experiences by the human test subjects of dementia although temporary was disconcerting to not only The Unified Coalition but family and friends of those transferred into the synthetic proxies. No one could be sure if the personality that they were familiar with had actually transferred. Although the Proxies responded with recognitions of their loved ones.

The Proxy's designs were chosen by their occupants and sometimes did not resemble the original human being. The design teams at C.I.T.A., could provide the human with their originals, but because these designs were at times circumvented, many general designs were offered in which the human could choose from assorted selections.

Whether the Chimpanzees would have preferred to die would continue to be an unknown. As for the humans that had become Proxies, there was little doubt that they wished to continue amongst the living.

A large indoor jungle was built to harbor the Chimp Proxies, and they appeared to flourish, quickly attaining the intelligence of a human child between the ages of three to five. It was theorized that these apes did not increase toward full human intelligence because their neuronal transference of information was fixed to that of a Chimpanzee's ability. They did not harbor the differential chromosome necessary.

Nano-neurons were in their infancy and Drake preferred to use embryonic cells. There had been an established history of these cells replacing neurons and circumventing dementia, it seemed the logical choice. Over time however, the synthetic neurons or nano-neurons somehow developed the ability of self-replication as biological neurons were as a matter of fact capable. They appeared to be learning or absorbing these abilities from their biological partners. Drake and Wyatt were unsure how such was happening, but they couldn't be happier with these results, and the Unified Coalition announced that enough of the transferences were occurring in order to call them a success. The Proxy program had the full approvals and support of the U.C., and would continue as the alternative to death. We could now look forward to circumventing death and ultimately, quite possibly, never dying.

The Unified Coalition approved just about anything that the Carons requested, and the only requirement for human test subjects was that they indeed had no other alternative other than death.

The initial attempts were sketchy as the human personality didn't seem to completely transfer, or at least appear to be the same. Some groups formed in protests.

These new bodies were in fact, different, which was theorized to possibly be at least part of the reason. Drake and Wyatt continued their pursuits, and Wyatt's lectures and classes were filled to their capacities, although protests appeared to be increasing.

CHAPTER 13: BEST FRIENDS

Whatever you do, or dream you can, begin it. Boldness has genius and power and magic in it.

Johann Wolfgang von Goethe

July in Barcelona was exceptionally hot for even this time of year. Gone seemed to be the relative perfection of balmy summer days. Temperatures were now easily comparable to sand deserts on average. Visual heat waves rose from what was left of the white sand beaches, creating mirages of ponds along the skimmed shoreline.   The rising of the oceans caused most coastal areas around the world to be covered. Even so, people flocked to the shores in hope that the shimmering blue water would deliver relief.

Global warming was indeed proving global, and most park areas now offered varied canopies of shade and misting devices, else the danger of increased amounts of heatstroke, exhaustion and other related illness. The older population appeared most affected, and corporate hospitals built wings dedicated to those suffering specifically from the heat, at a cost. Many of these businesses accepted whatever financial securities seniors had available, to the point of exhausting the currency flow and deepening governmental debt. This increased the demand for a different system of governments throughout the planet.

More and more people were flocking to the newly built underground homes of Spain. Having been constructed to avoid fallout these homes were built under portions of the ocean, and accessed through tunnels so large as to include vehicle lanes. These homes were generally called "dug outs" as they were dug out of the rocks and soil. New ideas toward saving our human population would be necessary if we hoped to continue as a race on this planet.

During this period, two women that would eventually become substantial to the change toward this vision, were but young hormonal adults with minds toward not only the opposite sex but an almost opposite location to further their education and adventure. Although they were determined to excel in their careers, little did they know how such choices would prove to be essential.

(Dialogues between Noriko and Dominique were in Spanish but is translated to English.) Dom grew restless; "This may be our last summer...come on Nori it's early enough to get a good spot!"

Noriko Matayoshi was a foreign exchange student to Spain. She along with a couple of others from her country of what was called Japan were placed under the care of the D' Angelis. During grammar school. Dom and Nori formed an immediate bond as Dom taught Nori Spanish and Nori reciprocated by teaching Dom Japanese.

Both girls knew English to some extent and practiced through school programs. Eventually Noriko's parents would move to Spain mostly due to Noriko's refusal to return to Japan. She had developed an affinity for Spain and would remain until she and Dom completed High School. Both were accepted to Berkeley University of California in the United States of America as foreign exchange students, and both would major in Architectural Design. During this period Berkeley was being relocated further inland and within the beginnings of bio-doming. Eventually such facilities would be updated with the increase of artificial intelligence.

Dom was growing impatient; "What in the world is taking you so long, Mija!" Dom questioned as she entered Nori's room. Noticing that Nori was wiping away tears, Dom's face took on concern; "What's wrong? It's beautiful outside...look! Hot, but the ocean will feel so good!" Dom walked over to Nori's window.

Noriko nodded slightly, "Yeah, I know... we'll be thousands of miles away, I'm going to miss it...aren't you?"

Dom nodded in agreement; "Sure I will, but we'll be on an adventure! Besides from what I've heard, California still has a few beach areas left. Don't worry...it'll be exciting, you'll see!" Dom smiled and tilted her head brushing her long black hair to one side with her fingers, then flipping it back. Standing erect and appearing firm in order to get Nori's attention, she stated; "Now come on before it gets too crowded!"

Nori nodded and dashed out the door laughing. Dom, realizing the dupe, shook her head, then smiled and ran in pursuit.

As Dom and Nori slipped off their robes, revealing bikinis, that left almost nothing to the imagination, they drew the attentions of several, young and old. Nori whispered to Dom; "I wonder what the men are like in the United States?"

Dom shook her beach mat open under the canopy; "Oh, I would guess that they're similar anywhere in the world. They're just convinced to behave differently depending on their cultures, but basically the same under it all." Dom grinned at the innuendo, as she gently placed the mat on the sand, attempting to divert the grains of seashells she removed her sandals on the slightly cooler area.

Nori looked out at the ocean: "They say everything is bigger in the United States."

Dom laughed; "Oolahlah, I hope so!"

Nori looked at Dom curiously and laughed once she realized her innuendo.

Nori shook her mat out and placed it next to Dom's, still standing, she grinned as she looked into Dom's eyes, and immediately Dom knew what that meant. Nori turned and ran for the ocean screaming as Dom chased her in a game of tag. Once Nori hit the first cooling wave, she screamed, and Dom was right behind her pushing her further into the waves. Nori came up gagging and Dom screamed attempting to get out of the shallow current. Nori dove forward and grabbed Dom's leg tripping her and making her sink under the water. Dom surfaced quickly screaming. Both girls continued in laughter as they spit the remaining water out from their mouths. Eventually they walked in deeper until the water was above their waists, washing their hair back in the lessened salty current and coming to the surface; quickly adjusting their tops and making sure the bottoms of their bikinis were still on.

Weather patterns were noticeably unusual throughout the planet, and many of the beach areas were a thing of the past. Spain's expansive beaches fell mostly under water now and sand was brought in attempts to rebuild what was lost. Some attempts at replacing the ocean's salt contents seemed hopelessly losing. The evaporation rates were increasing steadily and cloud cover was off setting temperatures in some areas, creating colder than normal patches, while other areas burned under longer durations of drought and heat. Many were experiencing extensive winters with snow falls moving into June in places like the United States.

Climate scientists worried that global warming would bring on a next ice age, but for the moment most of the planet seemed to be in a slow cooker. We appeared to have steered the planet into either an ice age or another Venus. It was speculated that once most of the Antarctic and Arctic melted, the world would be covered in clouds and the temperatures would begin to drastically fall thereby creating a snowball effect or what was called an "ice age." These choices were of course unacceptable toward our survivals, therefore the eventual formulation of a unified world government was considered essential. We would have to force one another to take this seriously by establishing laws thereby forcing behaviors.

Dominique and Noriko, as eventual Environmental Architects (unaware at this time of Universe City), dreamed of building the first biosphere capable of encompassing large regional populations in aesthetic environments that would entrance humans to remain within its walls.

Dom and Noriko hoped to work with space engineers incorporating their ideas toward space ships. They had gone to the extent of drawing up several possible designs; the first being a rocket launching system through an enormous pillar which the city would be located within and above. The cities would rise slowly by several space elevators, and launch once the proper altitude was reached. This of course would take an estimated six to eight weeks at best; we needed time for such launches and the ability to adjust to possible complications or obstacles if or when discovered. Once the cities were launched, they would operate completely self-reliant, supplying all resources necessary to the inhabitants. These designs were well thought out, and so impressively so, that NASA had actually taken their designs seriously. Dom and Nori were given hefty scholarships to the University of California at Berkeley, by a handful of philanthropists that were in support of not only environmental ideas, but ideas that could possibly save their own race.

CHAPTER 14: KALLIPOLIS

I know of no safe depository of the ultimate powers of the society but the people themselves; and if we think them not enlightened enough to exercise their control with a wholesome discretion, the remedy is not to take it from them, but to inform their discretion by education.

Thomas Jefferson

Mister Richards along with several other groups mostly now consisting of Universe City graduates, continued their missions of finding orphans for the future centralized Universe Cities planned for Earth's cities.

Previously, Universe City had grown to encompass eighteen miles, and nearly half of that sum being underground, with continuous construction toward a possible thirty, until disaster struck.

The Unified Coalition of the world governments currently consisted mainly of Universe City graduates and these efforts were fully supported. The graduates that departed Universe City for their original homelands, soon took positions of government, and worked their way up to levels of greater leadership. Many became the rulers of their countries. The world began to unify in an ultimate mission of peace, although forces seemed to play out in a battle of objectives as few terrorists' cells remained to hold fast.

 Schools and education systems across the planet began to incorporate the philosophies of Universe City. The majority of adults began to truly listen to and appreciate the opinions and suggestions of children. Our focus became our future in more ways than one, as children were seen as our greatest resource toward survival as a specie, and laws all over the world were instigated toward their protection.

There was however those that through whatever reasoning refused to consider gestures towards peace and they continued to instigate anger amongst their extremists groups. The Unified Coalitions popularity was soon to overcome even the child within such mindsets.

The drone wars would eventually come to a peak as cells were defeated by mostly the Final Two Forty with the Top Forty using their expertise with the smallest of drones the size of insects. Some even started calling the war...the swarm war.

Blade had graduated a few years before Jaiobian and was a commander in the underground world of war. He was but a handful of supervisors that could take control of the Top Forty Gamers brain linked helmets if necessary. With the few remaining cells now focusing on our underground population, vigilance was essential toward survival. Instructing his Gamers to lay down a fire of bees which carried enough dosage of tranquilizers to knock out almost any full grown adult male, was Blades personal preference. Perhaps because of his childhood, his affinity appeared to be to spare their lives, and perhaps steer their minds toward eventual benefit. If it were not for Universe City, he imagined his path to be a similar one to that of the cells. His techniques allowed our AIs to gather prisoners and our intelligence officers to obtain information in regard to future strategic movement. These prisoners were taken to our Mars prison facility.

Modern medical science was now capable of saving many. Recent and apparently successful transferences of much of the mind into mechanisms that were now being called Proxies looked promising. Doctor Drake Caron had pioneered this new life extension science and many humans volunteered as guinea pigs. The choices opposing death were to either transfer or be frozen cytogenetically.

The assumption was that these experiences were fully located in the brain, as we got closer to chemical exchanges and reproductions. Although this was Drake's and Wyatt's specialties they were perhaps more open minded because of Vicky's experience, however skeptical.

Many began to question their own beliefs their own chemical compositions as the planet appeared to deteriorate to a point of no return. Hope was turned toward the sciences as we initialized the process of finding our code, our numbers, our makeup to the point at which we could perhaps reprogram our systems toward benefit of the symbiosis of our planet.

Our thoughts were of despair as even the terra-farms on Mars that were built to replenish resources away from the war would not be enough to sustain the population much longer.

The world unified toward the pursuit of knowledge trying desperately to save our species. By epistemological interpretation, science and knowledge were one and the same. Therefore it was through the sciences and philosophies, that we searched and understood the meaning of life. In accordance with our history we used violence in order to reach many of our objectives. Our myths represented that violence and projected toward the varied after life scenarios. Heavens were a return to ignorance and hells were the punishments for lack thereof. Neither was any longer acceptable, and the human race began its evolution toward its next step.

Universe City was credited with showing many of us how to reprogram ourselves toward the benefit of not only our species but the symbioses that existed amongst the many surviving species.

Drake was asked to join in the attempts to replenish the Earth, and soon drones were diverted to seed the soil and clean the atmosphere through filtrates. Some tried to convince Drake to initiate the now called "AIO program" into the drones, but Drake refused, preferring to keep artificial intelligence focused on the eventual building of cities in which humans could be somewhat contained and content.

It appeared that some were indeed content in the biospheres on Mars. Perhaps it was that they were safer there than here from the war, as one never knew when or if a smaller drone sometimes the size of an insect was lingering near their vicinities. These drones could link up or create an explosive chain reaction of several other mini-drones. Detecting them required complicated systems that sometimes failed if our espionage information was not sufficient.

The paranoia increased, and underground security became stringent and tactics were not always appreciated. Scanners were commonplace and motion lasers and detectors were constantly in use. We became almost callous to these nuisances for the most part, but they did take their toll on much of the human psyche. Stress related bodily breakdowns became the number two cause of death, the war being the first.

Drake and Wyatt took an interest toward incorporating AIOs into personal vehicle transportation. AIOs could react significantly quicker than the human, and vehicle accidents could theoretically be nullified. The older classical computer models worked for the most part, but not well enough to where they could be fully trusted. Drake and Wyatt came up with several prototypes of vehicle brains, even to the extent that the human could sit in the back seat and enjoy the ride to their destination. That along with the building of further AIOs for construction increased the Caron's wealth significantly and Universe City continued to expand its population of orphans because of such generosities.

However, currency would be our weakness as societies continued to be economically challenged, and resources thinned. The wealthy were increasingly being despised and the Carons discussed new possibilities toward a government, the government that Universe City was in preparation.

On the technological front, there was hope as rudimentary generators and replicators were showing great promise; as we began to produce some soy based nourishment bars filled with all the necessary nutrients to keep humans alive. Mars was the main producer with AIOs working around the clock. The AIOs now numbering nearly a hundred and fifty thousand and producing their own operators were soon to finish the first twenty four fully self-functioning biospheres specifically designed for vertical farming on Earth.

These biospheres became so successful in production that many more were under construction around the world. The Unified Coalition supported C.I.T.A. (Caron Institute of Technological Advancement) completely and soon vertical farming under domed atmospheres was seen throughout the planet.

Humans began to leave their underground domains and some invested in their own personal above ground biospheres. However, the best or most affordable that we could do at this time for the dome construction materials was to enforce the polyurethane with thickly meshed fiber glass, supported by nickel titanium, aluminum and some rebar beams .

These materials were also used in the vertical farms on Mars and two of the small mostly underground cities. The construction on Mars however did include the latest black nano-shield outside the structures, which were initiated during storms and covered much of the domes underground. The translucent graphene and nanotubes would follow years later.

A few thousand AIOs had been transported to Mars for just such expansion. AIOs required no resources other than electromagnetic charge, of which they had been programmed to recharge themselves when it became necessary. A large percentage of the Mar's biodomes remained underground, and sunlight was brought in by sun tubes on Mars as well as the biodomes on Earth.

We did not however take into account the rest of the planet's symbioses, and soon we realized that the planet's other life forms were not regenerating as hoped. The vastness of devastation throughout the world due to the unmanned drones was obvious, and human's mindsets had changed too little in not enough time. It remained a conundrum how we could bring back life to Earth. Replanting by drone and AIO attempts were indeed beneficial, but so much had been destroyed that The Unified Coalition was overwhelmed, and we could not control the extreme weather patterns that followed, which seemed to conquer much of our efforts.

The world became a scary place as we tried desperately to rebuild our planet. It turned on us, and threatened to destroy us by not only the change in weather, but the continued collapse of Earth's magnetosphere. We however would prove that we were a stubborn race, and we would persevere toward saving the only true home we knew.

CHAPTER 15: Hormones

We were there at the perfect time, she said. I made it the perfect time, he chimed. Fated? She waited. No, calculated. You baited!

C. A. Solis

The University as were all other buildings was now mostly underground as people walked from area to area. It was considered safer to be underground as not only was the atmosphere no longer enough to protect life from harmful UV radiation but was highly dangerous from the nuclear attacks. Also the fact that drone technologies had long since fallen into combatant hands, so were not limited to the western allies. We continued to make drones, now specifically programmed to find other drones, and to detect viruses. Any attempt to divert programs was met with self-destruction.   The war would continue for another forty seven years.

Transport was limited to the older computer run vehicles with some of Drake's prototypes being incorporated into the underground transit system. Eventually Drake's designs would overtake the older models.

Dominique D' Angeli would very quickly catch the eye of Doctor Wyatt Caron as she walked to her classes each day.   Each time his eyes met hers his heart raced and he found that he couldn't breathe properly. He'd look away quickly perhaps defensively. He found this reaction fascinating as it had never before happened, or at least to this degree as far as he could remember. In his observer mode... he followed his physical responses.   He'd shake his head every time he took notice of her, as he considered himself much too busy to succumb to such tribal influence. Yet, she did indeed captivate him, so much so, that he found that what he could remember of any of his dreams, those fragments of his imagination, belonged to her. She flowed with ease through his thoughts and apparently his subconscious.

She was boisterous and that somewhat scared him as well, but her confidence was perhaps what most intrigued him. She was amongst his own opinionated probabilities to be the most beautiful woman he had ever seen. Was this but a shallow interest, he pondered. Long black hair of midnight with piercing green eyes and high cheek bones, and best of all...she was intelligent, very smart, three years older than himself, he had made sure to check with the faculty. She was top in her architectural graduate classes, and her designs toward environmental dome cities fascinated Wyatt, unknown to Dominique.

But how could he get the interest of an older woman, he pondered. What if she was finely tuned in the art of love making, to his far less untested skills? What if Don Juan himself would have difficulty matching her class? His face took on grimace and he shook his head in thought. "Wyatt, Wyatt, Wyatt...what were you thinking...she'd never be interested in you."

As Wyatt self-analyzed he realized that he should in fact be looking for a mate, and it was important to him that who he chose would carry over beneficial traits to any of their offspring. Intelligence wasn't guaranteed, but he hoped to increase its probabilities. "That's it, I should approach this, uh, her, with a confidence that mammals have when a possible mate is found. Just strut my plume and hope she notices, uh, I guess. I know, when in doubt, ask Leo!" Wyatt immediately called his best friend and confidant Leonard Cameron Starr.

"Are you being serious, Wyatt? You're a Caron! Of course she not only knows who you are, but I'd bet that you're at the top of her list of hopefuls."

"You really think so?" Wyatt appeared dumbfounded. Leo shook his head.

Wyatt was world renown. He was a Caron after all, and considered the top available bachelor in nearly every monthly. During this period, his image long with updates on his activities were featured not only in science periodicals, but practically every popular adult and teen websites and magazines available.

Hopefuls on occasion nearly fell over as he passed, and many tried varied means in order to draw his attention. Wyatt however was a bit aloof and incredibly focused on diverse branches of physics. It appeared to be what kept his hormones perhaps from controlling his behaviors.

Dominique however was on a mission to catch the top fish and her bait was undeniably being distributed in Wyatt's stream and selectively his stream.

Leo was highly intuitive on many occasions to where the Caron's relied on his feedback. A hunch from Leo was considered an educated guess which always seemed to be correct. He'd commonly comment on whomever and whatever was attempts at hooking Wyatt. However, he held back any comments when it came to Dominique. In fact, Leo appeared unusually supportive when it came to Dominique.

She was about to take advantage of Wyatt's occasional primate regression, and on this occasion, as he stopped and for that moment focused on her, she told him that if he liked what he was looking at then he should ask her out, to which he was stunned. Leo's intuition was once again spot on. Wyatt's voice screeched and he quickly cleared his throat. "Uh, hmm, excuse me?" he responded, his voice squeaked slightly, indicating his inability to correct his condition. Embarrassed he began to walk away.

In her Spanish accent Dominique called out, "Just you wait a minute you, Mister Wyatt Caron!"

"Do we know one another?" Wyatt stopped and turned in defensive response.

In nearly a whisper, Dominique responded, "Everyone they knows the Great Wyatt Caron."

His forehead wrinkled as he was yet again caught off guard in an attempt to not only hear her voice, but to that which he felt a sense of security toward her regressive perhaps submissive behavior. "Ah, I see." Wyatt straightened his books and papers being held. He took a position of control, standing straight, his shoulders back.

She composed herself and smiled widely then announced, "My name is Dominique. How do you do?" She extended her hand shake. He moved the pile to one arm and reciprocated by shaking her hand and smiling back responding with a nervous "Nice to meet you, uh...Dominique." He took in a deep breath toward regaining control, "Would that accent be the Barcelona region?" She nodded, "Yes! And please call me, Dom. All my friends call me Dom." Wyatt nodded, "Okay Dom." An awkward pause followed.

She looked into his eyes, "So, are you gonna ask me out... or what?" He was taken a bit back by her now obvious boldness, "Uh, I guess so, actually... I'm very busy." She shook her head, "Okay, adios, nice to have met you." She turned and began to walk away.

Wyatt was stunned and somewhat confused. His opportunity was here and he foolishly was brushing it away! "Uh, wait! Whaaat, what, what about tonight?" Wyatt realized that this was perhaps too soon and possibly not appropriate and his face took on embarrassment.

"Sure! No problem...perfecto!" Dominique responded. She walked up to him and wrote down her phone number on top of Wyatt's mound of papers that he was carrying.

"Uh, hope I don't try to calculate those into that formula." Wyatt grinned and chuckled at his own joke." Dominique giggled…yes…she actually found Wyatt's humor quite entertaining. Wyatt smiled, "Would seven o'clock be okay? I should be finished for the day." he muttered.

Dom smiled and nodded, then put her hand over his. He wondered if this new feeling was in fact the static electrical charge generated. He looked at her with intense interest. She continued, "Seven? Of course. Don't you be late, I don't wait for anyone, not even the Great Wyatt Caron. Okay, well maybe I'll wait for you a little bit, but try, okay?" She squeezed his hand, winked and grinned.

Bemused and much more relaxed, he once again straightened out his posture, cleared his throat and confidently responded, "I'll be there promptly at seven." He pointed to her phone number as if it were an address. "That's near here right outside campus."

"Si. Oops, sorry." She scribbled her address next to her phone number, and drew a little map to her dorm. "My roommate and I are foreign exchange students. Don't worry Professor, I'm a graduate and working on my dissertation. Noriko and me, we're are a working on Biodome designs, and should be able to prove much better environmental possibility toward our designs' incorporation. Not to worry, it is in no way associated with your studies. Not, uh, no conflicts of interest whatsoever." She ran her hand down his wrist to his hand, and she could feel his rapid heartbeats. Their fingers intertwined for a moment, and she smiled as she rose on her toes giving him a soft kiss on his cheek. "See you tonight", she whispered as she walked away.

Wyatt stood there for a moment smiling as his heart continued to race. He now knew that it wasn't static at all. These feelings were indeed electrical, and apparently meeting no states of resistances. He watched her walk away, realizing that he was focused on her sway, he shook his head bewildered of his feelings. He felt silly although exhilarated, as he dropped what he was carrying to the ground. He quickly picked up the books and gathered the papers. Looking up to see her once more, but she was gone.

Wyatt's best friend Professor Leonardo Cameron Star supported Wyatt with endless patience as he listened to Wyatt for the first time ramble endlessly about a subject which appeared beyond the full grasp of physics.

Leo held a Literature position at the University as well as Anthropology. It was also known that he held History and Sociology degrees. By his students he was voted as the "Best Professor All Around." No one knew exactly how many degrees Leo possessed or that he would admit to, but he could hold a conversation with anyone no matter their background. Some students called him "Leobrarian" indicating that he was likened to a library, the library more than the librarian, that is. The Carons had welcomed him fully into their family. Drake and his son Wyatt had found him to be possibly, to quote;" one of the nicest people" they had ever met. "Soft spoken with an intelligence and wisdom that seemed otherworldly." He was a large part of Victoria's Venue as the organizer for the Orphan round up. He personally met each and every sociologist and psychiatrist before they embarked on their missions. He had apparently traveled extensively throughout the planet, and was essential to the organization.

"I have a feeling that she's perfect, Wyatt." Leo smiled. "Leo, I'm supposed to meet with Professor Caldwell tonight. Can you cover for me?" "Sure, sure, not a problem. What's the subject that you and Ron are on?" "Ah, he just needs a little more confidence on his research. Quantum Mechanics is messing with his project again, always seems to kick his ass when it comes to bringing down the macro to the mesoscopic. You know how that goes. He just needs a little guidance with the unity of uncertainty…electron liquidity." "Yeah. Okay, I'll try. It's been awhile since I tutored you. But it should come back to me."

Wyatt handed him a few sheets of papers, "Here's my research on his dilemma. He should find some interest in these, particularly this part here." Leo read the area in which Wyatt pointed to and then tapped, "Alright, I understand, it's a date. I'll take him out for some pizza and a glass of Carona!" Leo smirked, then glanced through the rest of Wyatt's research papers, "Wow Wyatt, you always amaze me, so much like your father and his father before him."

Wyatt looked surprised, "Really? You knew my grandfather? You never told me. You're, you're not that old, are you? Wyatt with anticipation awaited Leo's response, however Leo was reading through some of his research. Wyatt continued, "I always feel like I'm one step behind, so, uh, when did you meet my grandfather?" "Behind?" Leo shook his head, "Uh, huh, you're doing fine. I read about him." Leo continued to read through Wyatt's research. Wyatt paused then nodded, seemingly satisfied with Leo's response.

Dominique's best friend Noriko appeared to be on the same boat of her friends' apparent temporary insanity. Dom took deep breaths in an attempt to build up her courage and plan toward Wyatt for the night. She was beyond interested, she was tremendously intrigued. One might say obsessed. Everyone knew who the Carons were, but Wyatt was by many seen as representative of the hope for this world. There were many other geniuses, but Dom was amazed that someone so young was teaching, and teaching the higher mathematics of quantum mechanics, and lecturing on world changing innovations. His heart, his essence of qualia was what she wanted more than anything. She wanted his world, as seen through his eyes…she wanted it all, all of him.

Yes! She too wanted to change the world! Help in making the world a better place for all! Sure he was a Caron, and everyone knew who the Carons were! (Gulp) She swallowed and shook her head at her own silliness. She was never interested in wealth or status! She detested intellectual snobs! He was an Oligarch! Was he? No…he was different. He and his family were loved by most around the world. Heck, they started Universe City! Universe City, yes, the place that was her motivation toward architecture and saving the human race! She truly wanted to dedicate her life to bringing back the beauty of the world through architecture. This was more, much more as he was consuming her every thought which had never happened to her with any other. He just couldn't be perfect, could he? Her mind was racing as fast as her heart, and she felt both exhilarated and exhausted.

They hadn't yet experienced their first date, but she was hooked, and was driving Noriko over the edge of her own sanity. Noriko screamed, "If you don't slam dunk this man, I will!" Dom laughed and high fived Noriko's suggestion, then fell back on her bed screaming until she sighed from her use of what seemed like every ounce of energy that she possessed.

Noriko helped Dom up, and they both pick out her skin forming blue dress for their evening. Looking at one another, they both nodded and grinned in affirmation that this was the perfect weapon against any defense he may have. Dom slipped it on and pulled it down slightly over her knees. It wasn't hard to guess what was under the form, and she didn't want to show any more than perhaps was necessary.

 "Did you know that his favorite color is aquamarine?" Noriko questioned as she combed Dom's long black hair. Dom's face took on surprise in the mirrored reflection. "Where did you read that?" She asked while looking at Noriko's reflection.   "I didn't read it anywhere. Everyone knows that his best friend is Professor Star. You know…"Leobrarian", Noriko's gestured the quote then continued, "Well, I asked him a few questions about Wyatt." Dom was fully focused, but Noriko became silent.

Dom rolled her eyes and slightly nudged Nori, "Uh, come on Nori! Ay yai yai, it's getting late and he'll be here soon!" Noriko started to laugh and Dom picked up a small decorative pillow and aimed it toward Nori. "Okay, okay…those beads can cause damage! Umm…ooh…and I think that Wyatt is a virgin, oh yes, ay yai yai! Ha!" Noriko smiled as she placed her chin on top of Dom's head. Dom eyes opened widely, and Noriko continued, "Or at least he has never had a girlfriend to Professor Star's knowledge." Dom threw the pillow at her. "Hey, that's not fair, ouch! I told you everything that I know!" Dominique sat back in thought, then grinned. She shook her head realizing that she was removed from what should have been common sense.   "Nori! I am so sorry! Are you okay? What's wrong with me?" She embraced Nori tightly.

Nori gasped as her arms were pressed to her sides. "It's alright. I understand. I'd be just as nervous and messed up if it was me. Okay. You can release me anytime now…I need to breathe on occasion!"

Dom released her embrace and they both placed their foreheads against the other and laughed as they yelled "Cyclops!" in unison. They had done such throughout their years together whenever they found themselves without sufficient words. Somehow seeing one large eye (as their own eyes crossed) in the middle of their foreheads, caused them to laugh uncontrollably no matter the circumstance.

Wyatt's' and Dominique's first date took place at his apartment in which he had Juan cook a fine dining experience. His flat was located near his parent's home on the same grounds, mostly for security reasons. Wyatt would not only lose his virginity that night, but fall deeply in love. He could not go a day without being with her, and she him. Dominique taught him the many benefits of sensuality, and as always, Wyatt learned quickly.

Because of his relationship with Dom, he almost immediately moved his teachings to the headquarters of C.I.T.A. as Dom finished her PhD. This allowed him a bit more freedom at his labs, and avoided any conflicts at the University. Students were always welcomed, and would meet at C.I.T.A's lecture rooms in order to hear not only Wyatt's briefings, but many of the other Scientists.

The Carons were strong advocates toward the sharing of information as long as such was not considered detrimentally capable. Such philosophical holdings had conflicted with their desire towards Universe City's initial secrecy, and now they realized that the contributors including themselves should perhaps had been less clandestine toward its existence. However, that was in retrospect, and during that period, what had mattered most was the safety of not only the children, but of everyone involved.

They would become engaged less than two months later to both their parent's surprise. The D' Angelies were however well aware of the Caron's reputation and in fact had been contributors toward their philanthropic organizations. They were beyond thrilled. They were not in the Caron's financial arena, but they were proficient organizers, and Dominique and Noriko were well indoctrinated in regard to hard work with the focus on generosity toward others.

Dominique's heart toward philanthropy won the Carons over almost immediately. Drake and Vicky couldn't be more pleased, and hinted that a grandchild would be more than welcomed at any time, and had wondered if such were the situation. Wyatt and Dominique assured everyone that no such condition existed. They however did plan on having children as soon as possible.

The Carons fell in love with Dominique almost as much as their son. She made their son complete and most of all, happy. Wyatt's father Drake with his usual humor jested how she was "much more beautiful than Florence", "Wyatt's first sexual interest." Wyatt noticeably cringed a bit, then informed Dom that he would explain his father's dry humor later.

"Don't worry mi amore, mi madre would sometimes embarrass me by screaming, "culata abrazadera!"

Wyatt looked on curiously, "butt clamp?"

... ai yai yai, si, si, yes,...she would ran, uh, ai, sorry, she would run, run after me trying to grab my buttocks. If she did caught, catch me, she would hang on, ouch! Sooo, so, sooo embarrassing!" Dominique put her hands over her mouth in embarrassment.

The Carons including Wyatt laughed. Vicky stated, "Your mother sounds like fun!" She grinned. "We can't wait to meet your family!" Vicky smiled.

"Si, you and she and the family should get along very good." Dom smiled at Vicky, and Vicky reciprocated nodding." "I'm sure that we will." Vicky stated. Dominique continued, "Our family is small. Many have died because of the war."

Vicky eyes saddened, "Yes, as is ours. Hopefully it will be ending soon. It's said that the gamers have cleaned out most of the cells."

Dom nodded, then changed the subject by questioning Wyatt, "So, who is this Flo, Florence? I thought that I was your first and only?"

Wyatt's face flushed red as Drake and Vicky looked on a bit surprised. Wyatt whispered to Dom, "Uh, I'll explain later."

Drake purposely interrupted, "so, when is the date?" Wyatt appeared relieved that his father had changed the subject.

CHAPTER 16: UNIVERSE WEDDING

I asked him what kind of a wedding he wished for; He said one that would make me his wife.

Author Unknown

The two families were surprised by the date of the wedding as it had been set for a month following Wyatt's and Dominique's announced engagements.

They had first thought of eloping, but worried that their families may find such insulting...so they went ahead and gave a one month's notice, with the wish that Universe City be included. Dominique had deeply fallen in love with the atmosphere and of course, the residents. She was amongst the fortunate adults that was allowed to visit when she pleased. Universe City to Dom was paradise. She couldn't think of a more glorious place to be married than amongst so many children.

They decided on a very short service of not more than fifteen minutes due to many of the attention spans of the majority of residents. There would be celebration with plenty of games and a cake made of assorted colored lollypops each in a pink and blue on frosted chocolate, vanilla and strawberry cupcakes. Her love of children was apparent and she wanted several or as many as she and Wyatt were capable.

The Caron's and Angelis' had little problem in getting it organized as The Coalition took control and the event was made ready in the relatively short time. The biggest problem that arose was the attendance as so many wanted to attend. Because the children of Universe City were the main attendants, most around the world accepted that viewing the nuptial exchanges through various video relays may be more advantageous or enjoyable as not everyone had the patience for over a million children.

The time passed quickly...and the centralized garden or as the children liked to call it "The Enchanted Forest" was filled with children sitting everywhere, on the trees, on the ground, lining the small river that flowed throughout the city to the various swimming pools.

Wyatt and Dominique would be on a very long and large wooden raft as their promises were being spoken. The raft was being rowed on the river at a steady pace by Juan. Juan was instructed to take the raft down the river in fifteen minutes, so the pace would be quite fast, but steady enough for the occupants. The immediate family members sat on attached chairs. The raft was carved by AIOs to look similar to the Kon-tiki with elaborated glittered tikis at each corner and a sail adorned in the style of Polynesian Tapa cloth. The glitter was mostly for the children as they awed over the site of the ship.

It was symbolic of Wyatt's love of languages and of his mother's love of people and her field of anthropology.   Dominique felt so much at home with the Carons that she hoped that her interest in architecture could and would be collaborated with their philanthropic efforts and help in the development of a future world in which life would thrive. She was certain that with the help of her new family, our species could be saved. She was beyond happy, she was overflowing with hope, and with that, excitement toward her and her about to be husband's future.

Dominique's gown was simple and elegant. It was aquamarine (Wyatt's favorite color) with white lace speckled silver and crystal flowers that sparkled as the sunlight shown downward through the top of the main dome and sun tubes near the river. Her long black hair picked up the slight atmospheric breeze of Universe City, and the blue green gossamer chiffon that lightly layered her gown rose like wings gently flapping. As the raft moved down the river, her gown's train rose slightly, the chiffon moving like flocks of birds against the depths of sky. Children awed as she passed by, and Wyatt looked on as a pleased peacock disguised as penguin dressed in the traditional black tuxedo. He occasionally touched the corners of his eyes in order to absorb any overflow of his apparent emotion.

Dominique glanced over at him and as their eyes met, she and he would grin as if keeping a secret, the secret of lovers.

The moment arrived and they joined hands...and three Justices read in Unison accepting Dom's and Wyatt's "I do's." Then Dominique could be heard saying her vows, "I will love you always, mi amore. I will support you and be there in your most needed moments. I will grow old with you." She smiled, then winked and whispered "I'll try not to get too old."

Wyatt grinned and then responded, "I want to grow old with you, and look old with you and I will love you and only you for the rest of my life as a mate loves its mate. I will work toward making you happy...and try hard to have fun throughout our life together. I will fight for you and never let you feel afraid or threatened by anything, ever. I will never stop loving you." Tears flowed down Dominique's face and Wyatt placed the modest ring on her finger. She wished that she had said more...told him how he gave her hope...a gift more precious than anything she could have asked for. Her love for him made her eyes spill tears of happiness, and she uncontrollably smiled and laughed. He kissed her for such a long period the children's laughter turned to a loud "ewe!"...which caused the entire marital party to laugh.

She chose a ring that reminded her of the situation of the world. The diamonds were very small, actually chips held to the band of black graphite by several prongs. The shiny black background gave the appearance of a solid diamond band. She assured Wyatt that he could get her a larger diamond when they no longer had value, as nanotechnology was in the beginnings of producing much stronger materials such as graphene, and the hope was that soon it would come in varieties and be made translucent. There was little reason why it couldn't as diamond was simply a form of carbon, granted very compressed, as was the carbon based nanotube.

The children then cheered along with the laughter, and jumped with glee and some covered their eyes as Wyatt kissed Dominique again, dipping her backwards as he pretended to twist a non-existent mustache . "Brouhahaaaa!" He shouted. The sounds of a unified "ewe" were heard yet again, and Wyatt and Dominique broke out in laughter along with everyone at the event. Almost immediately the displays of fireworks were seen through the dome's ceiling, and the celebration began. Everyone cheered.

CHAPTER 17: DECISIONS

You cannot get through a single day without having an impact on the world around you. What you do makes a difference, and you have to decide what kind of difference you want to make.

Jane Goodall

Jaiobian had spent a few weeks away in several countries observing their governmental systems in order to deliver her observations that were required to graduate. By now many countries were run by graduates of Universe City...so in a way she was at times visiting with family. It was always joyous when the graduates met, as although they were needed in their countries, they missed their Universe City home and seeing another graduate, many times, brought back the flood of wondrous memories. Of being saved from horrific circumstances and then seeing what the world could be if only we could cooperate and look to one another as we would family, a family that benefitted us and that we eventually benefitted, and of and to which we loved and were loved.

The graduates were developed to make the world join in benefit. The worlds' rebellious terror cells nor the separation of classes would prove to never be able to tackle what Universe City had built. Strength unseen, indivisible and determined to overcome the human weakness of violence, these were the Philosopher Kings once dreamed of, and Jaiobian would soon be joining their ranks, and becoming the mother of her nation.

Jaiobian had been playing and graciously swaying to the music from the piano. She stood up as she noticed Mister Richards approaching. She knew what the coming discussion would encompass.

"Mister Richards, you do realize that I want to stay here permanently?"

"Of course I do Jai, but we didn't expect your country to request that you join their parliament. Heck Jai...you were voted in and you're not even there! You won by your reputation here. That's extraordinary! See how word gets around?" Mister Richards smiled proudly.

"I know it's a great honor and all, but... (Jaiobian looked up at the domed sky as a few birds flew toward the inner garden) it would be so difficult for me to imagine any other home that could compare to this, Mister Richards."

"I know Jai, and that's precisely why you have to go home, in my opinion. It's your purpose. You were brought here to do exactly what you will now do in your country. It's now up to you to make it like it is here. I've heard that much of the city is now underground as is the United States. You'll have plenty of support from the Alliance and I know that you can get the wealth to flow your way...most of the philanthropy for Universe City was from those that believe in this philosophy."   "Yup, I suppose that you're correct...(Jaiobian hit a few of the lower keys of the piano in unison) oh, I know that you're correct Mister Richards, of course you are...I'll go, I have to, if that's what it takes to turn this world around, I'll go and do whatever I can in whatever amount I can."

Mister Richards persisted, "You're going to make the best kind of leader, Jai. The kind of leader that cares more about your people than yourself or your position. I know that you wanted to give back here at Universe City and be a teacher, but I've observed you for so many years now, and you always focus on making sure that the others come first. So when I heard about the request, I knew that you were ready and that you are exactly what your country needs. A leader that's a teacher. Can't get much better than that." Mister Richards smiled.

"It's going to take a lot of time and work." Jaiobian shook her head in contemplation. "Not saying it can't be done, but I'll need some help."

"That's what Universe City is all about, Jai...help. Many of the world leaders are graduates; you should have all the help that you need...just ask. Perhaps you'll eventually be voted into the new Unified Coalition? It appears that the Alliance of Nations will eventually morph into the Unified Coalition. The Coalition sure did stimulate an unexpected grand unification of nations."(Mister Richards laughed) "I'd guess that it's coming together faster than anyone expected. It appears to be what the majority of this world wants."

Jaiobian nodded, "You're right, Mister Richards, as usual." She smiled largely. Paused in thought, then proceeded, "That is definitely my goal! If I can't stay here at Universe City than I'm bringing it there! Before you found the Crows, Mister Richards...I had thought that this Universe hated me somehow, but then something, perhaps what others might consider small made me realize that wasn't the case at all."

"What was that, Jai?"

"Well... I had found a potato that was sprouting in the trash behind Mister Gonzales restaurant. So, I took it and planted it in some of the soft soil near a really big boulder that the Crows used for shelter sometimes. We dug quite a cave under that thing. Anyway...so I planted this potato and I made sure to throw water on it occasionally, and sometimes I'd pull up the weeds around it and so forth. It grew a little, and at first I thought; sure enough...the Universe hates me...it's going to die. I killed it because I'm a jinx, you know, that sort of thing. But a few weeks later, Sheba was helping me control the weeds around the area, and she decides to dig it up and check how far its roots are going! And I began to freak out, but... then... she brings up at least a dozen potatoes! I mean, they were huge! Biggest potatoes that I had ever seen!" Jaiobian's corners of her mouth increased to a large smile as tears filled her eyes, and her voice trembled, "Get it, Mister Richards? The Universe didn't hate me after-all! This world is what we make of it." Jaiobian wiped her tears off her cheeks.

Mister Richards smiled and hugged Jaiobian, and then gave her a kiss on her forehead. "I understand, Jai, I understand. That's incredibly profound that a child of that age could gather that insight from such an event. I can't tell you how proud I am of you, Jai." Mister Richard's eyes filled with tears.

"Ironic isn't it Mister Richards...you're now the one telling me that I need to go back." Again. Jaiobian's eyes filled with tears and she wiped them off her cheeks quickly.

Mister Richards nodded, "But you see Jai...we just...we had to first remind you of how amazing you are, and then release you to the world."

Jaiobian hugged Mister Richards tightly as if saying goodbye.

"Hey... hey... hey, Missy...this is NOT goodbye! I'll be visiting you and I'm sure that I'll come with plenty of company. Sheba will be graduating in a few years, and I suspect that she'll pitch for a place on the parliament as well. Come to think of it...Blade and Craze are there." Mister Richards shook his head and grinned. "I'll probably always call Bobby and Carla by those nicknames. I wonder if they had anything to do with the request. Hmm...I just betcha!"

Jaiobian smiled, "does make me feel better that they're there. I hope that I can do as well as they have."

"My bet is that you'll do better." Mister Richards put up his hand for the high five that followed. "I'll try my best." Jaiobian nodded as she reciprocated.

"That's all that's expected, Jai, we know that you will...this place is a part of you now, and even though you're leaving it, it will never leave you." Mister Carla is the only one that I know right now on the board. Bob is the Mayor of our City, but he wants the governors' seat from what I've heard and I'd bet that he'll probably get it." Jaiobian nodded. Mr. Richards smiled and again kissed Jaiobian's forehead. "Will you allow me to escort you to The Coalition...they're waiting for your decision."

"So, they sent you huh... you're their secret weapon?" Jaiobian grinned.

Mister Richards patted Jaiobian on the back, "Nope...*YOU* are."

CHAPTER 18: CHANGE IS THE ONLY CONSTANT

Laws and institutions must go hand in hand with the general progress of the human mind.

Thomas Jefferson

Universe City which was located on an island, had massive walls built around its domes, in order to keep out the encroaching water moving into the forest that surrounded the city. It had been anticipated due to global warming, but no amount of preparation appeared to be sufficient.

The underground tunnel's walls were made of six overlapping two foot thick steel platelets   that could shift under several layers of a mixture of enhanced flexible concrete when necessary. No one knew how long this could hold back the impending threat of the ocean, but there was hope as graphene was slowly being incorporated into the newer construction materials.

Unfortunately, we didn't know it at the time, but due to perhaps conspiracy theories and propagandas spreading throughout societies, there were developing organizations, *terror cells,* bent on the destruction of Universe City.

Groups had developed that felt that children should be allowed entrance based on financial capabilities. Some tried unsuccessfully to purchase special entrance for their children. We weren't sure if these groups stimulated negative actions. The suspicion appeared to be that the mindset of human's impoverished, in this case the world's orphans would circumvent that of the wealthy. We seemed headed toward a world of equality as never before achieved. Many were uncomfortable with this possibility, and Universe City's only failure appeared to be its tremendous success.

Rumors had picked up steam about the apparent construction failure of Universe City. Some claiming that dome cities were not our solution, and that we perhaps needed to establish cities in outer space. We had successfully built a connection of domes on Mars, but mostly for prisoners and prison staff. A few hundred scientists were also assigned toward various studies and in separated domes. The initial space travelers during the later twenty first century had proven to us that Mars was not our answer to over population. Many died as a result of the Martian toxins, and death was not considered an answer by the newly forming governments which sought instead to unify human ingenuity.

Eight space stations were currently in use. It became imperative that we find a financial solution to domed space cities, in order to equalize availabilities toward the general populations. The space elevator was amongst many suggestions and would eventually be incorporated in order to transport materials.

Many amongst the world's wealthy, organized and financed campaigns against the efforts of Universe City. Quite surprisingly to others, however, scientists around the world also organized in unified efforts against these attempted oligarchies. The initial Coalition members along with many other governmental leaders eventually became the Center of Information or "COI" as they were called. They convinced governmental taxations systems to flow toward unified efforts, which began to bring together leaders from various countries. Many of these leaders were in fact graduates of Universe City. The Carons along with their friends and associates were amongst the COI, and soon the fully established Unified Coalition would incorporate the ideals of the COI. Pulling resources together, hundreds of thousands of research facilities were established, and millions were employed. The oligarchies tightly united, and some were rumored to have joined forces with several terrorists' cells. Increasingly, the governments taxed these corporations in order to stimulate employment in governmental research facilities. The oligarchies however found such attempts to be direct attacks. The worlds' oligarchies had grown accustomed to acquiring indentured servants in the form of employees. The COI offered a way out of debt by governmental release and settlement. Governments absorbed all public debts, and through applied general taxation and controlled interest entitlements, such debts were resolved.

Investigations proved that some of the damage to Universe City, was in fact caused by sabotage. This angered many around the world, as during this period there were several million children residing in the city. The walls were thought to be impenetrable. The city had been built to act as an island within an island, and now being reinforced with graphene wasn't sufficient. The possibility that water levels would rise was now happening, and leaks were increasing. By all reason Universe City should have withstood the ocean pressure and in fact semi-floated within the island as if the center of an atoll. However, it was leaking, heavily. Leaking and the ocean moved within the walls and under the city, and it slowly began to sink. The sabotage was linked partially to the design team that had purposely not continued the design to enclose the dome as a surrounding bubble.

A group of contractors that were amongst the many that originally designed and constructed the walls, were found and prosecuted. Many around the world demanded their deaths. It was decided that they would be sent to the Mars prison facility. The toxic soils of Mars and the vigilance of camera surveillances would be enough of a deterrent to keep them in the facility, and they by reason had the education enough to continue its upkeep. As well, they would be employed to further construct their own facilities. Staff was kept to a minimum, mostly for law enforcement to some degree, and would eventually be replaced by AIOs. The only returning spacecraft was for those scientists, assigned to Martian planetary studies, and the prison's staff, circled the planet. This spacecraft was only capable of operation once Earth Control Center initiated ignition and landing onto the Mars surface. Once boarded by scientists and/or staff, ECC would then control takeoff, flight, and return to Earth.

The Unified Coalition did not give death sentences. Life sentences however, did not offer parole. Thus such a crime that justified life was classified at the ratio of damage it did toward human ideals. These criminals were of course considered human, and could by reason create their own society, however, they would not be allowed amongst Earth's societies ever again. They were banished from Earth. The sabotage was considered so vile because of the amount of harm that could have occurred to Universe City's population which was mostly children. Children, during this time of a ravaged planet due to the ongoing wars, were considered our greatest asset, and harm to such was highly punishable. Mars consisted of all prisoners that were found to have harmed children.

The reasoning given by those responsible toward the sabotages', was that "Universe City is disrupting the natural destiny of the human being." They called themselves the "Doom Doers", and looked forward to the annihilation of all life. However unproven, it was rumored that they were financially supported by the united oligarchies. Mars would perhaps be their ideal...although life was found on Mars in the form of bacterium, which was being further explored by those scientists residing in the separated Bio-domes. These Doom Doers along with the other prisoners would be the labor force necessary on Mars.

CHAPTER 19: RESPONSIBILITY

For a successful technology, reality must take precedence over public relations, for Nature cannot be fooled.

Richard P. Feynman

At its beginnings, many of the leaders of the Unified Coalition were coming directly out of Universe City. The planet however was in dire need of these leaders yesterday. Our need of them was perhaps to bring us all to our senses, to save us from ourselves, and devote what we could to reestablishing life and that which supported it.

As our world suffered on...it became increasingly necessary for us to figure out ways to collectively save whatever biology existed, all of it, or as much as we could. The symbioses toward human survival was apparent as we came closer to possible extinction. Several species that we regarded as docile, some even as pets, were now running wild and scavenging what they could in order to survive. It had become obvious that something had to be done in saving what was left of the living creatures (including us) and plant life.

Even though humans had developed several varied replication and regeneration devices in order to create nourishment from the molecular level and possibly soon from the atomic level...we none too soon realized that information had a considerable linkage to all life forms and if such information could not be established, we could possibly be next with those on the extinct list alongside the billions of others.

Synthetic meats were a large success that was based on animal stem cells as we were weaned away from the animal itself. Although several trillions of stem cells were kept available for the replication devices...it became necessary on occasion to extract fresh stem cells in order to refresh the supply. Animal husbandry was transformed into a science in which animals were not harmed in the process of stem cell cultivations. They were in fact well fed and kept in organic environments, growing rapidly as humans continued to reestablish various biology.

Eventually only a very few animals would live amongst humans in the biospheres. Most were reestablished in the wild, however many required shelter from the increasingly harsh weather; human intervention by AIOs was necessary in constructing varied shelters and supplying nourishment. This was considered essential, and therefore several biospheres were approved and erected just for such purposes and maintained by scientific communities. These particular biospheres were representative of the variety of environments crucial to these life forms.

Jaiobian had been elected as Prime Minister of her nation as well as the highest government position in the world, "Spokesperson of the Unified Coalition", the UC. The world had not only become familiar with her story, but with her current accomplishments, and she was greatly admired. She had established work stations in the fashion of Universe City throughout her country, in which, anyone could be trained and educated for free; in any industry or talent. If you had the will, you were guaranteed a job once you were educated and/or trained, whether that be in industry or governmental. It also gave those that succeeded in their studies, an open forum toward elected positions.

She was well aware that her country had to keep the flow of currency moving or there would be economic collapse. A ceiling or cap was put on wealth and any corporate hording over five hundred million was put into the economy, unless the business funded areas of philanthropy. Governmental funds were also established toward specific public benefits.

She encouraged and promoted philanthropic efforts by honoring people with such objectives throughout the country. Statues were erected of outstanding Philanthropists and announcements of their achievements were widely publicized. Corporations were also awarded for their production and abilities to contribute.

One needn't be wealthy to be honored, as there were several that were honored organizing various benefits...such as the cleanup of the river that she grew up near. On occasion she took leave to return to that little town and the tributary that she grew up, which was now what looked like a new clean flowing river. She stood on the large bolder where Mister Richards spoke those harsh words as she searched for any treasure that might ease the pain. She smiled because there was no trash to be found, and she remembered when she first received the offer from Mister Richards. She reminisced and noticed a few crows that had survived the ongoing but lessening Drone War. Perhaps they survived because of their intelligence. She made sure that they would in particular be well fed. She ordered several bird feeders to be filled daily and placed in the woods. The children of the streets were few, and she encouraged them toward education, and who remained she employed and paid them to feed the birds.

She made sure to have the newly designed vertical farms done in the same manner of those at Universe City. Such was how Universe City produced more than enough food for its residents. Its first full bloom of crops was unfolding, and seeds and nuts were being produced in order that what was left of the other birds and small animals could not only survive but thrive and replenish growths.   Honey bees and humming birds were slowly returning, but artificial germination continued to be necessary for some time.

She made sure to organize planting committees in which several varied fruit trees were planted in every public park, and gardens were encouraged in areas designated.

During her first visit to Antonio's Bakery, she was greeted by almost the entire town cheering her name. She was after all one of their greatest achievements, their Nation's leader.

As she bit into the free cupcake that Antonio himself prepared for her with her favorite pink frosting...she imagined what her life may have been like if not for Mister Richards and Universe City. She thanked Antonio, her tongue bright pink, and she smiled as he nodded offering her whatever she wished. "This is fine Antonio...it's as good as I remember." She took a sip of milk, wiped the corners of her mouth with her own handkerchief in her suit pocket, and then she walked out the door that was now much larger than before, following its last renovation; perhaps because crime was considerably lower, nearly nonexistent, and she headed down toward the old fitness center.

People waved and cheered with delight as she looked from side to side, her security walking not only next to her but jogging in the front and back as well, and some moving into the crowds that were not only getting larger but many had begun to follow her entourage by running alongside. On occasion she'd stop and shake hands with a recognized face. Some had treated her badly when she was a child, but she held no grudges.

A young man waved from the third row in the crowd, "Jaiobian, hey! Remember me? Micah?" She smiled largely pushing through the crowd, they gave way as Micah picked up her petite frame in a hug. Her security stepped in, "It's fine, relax...he's a friend."   "You look amazing, Jai! Far from the little ragamuffin I knew as a kid. I have to admit, I was kind of scared of you and your gang back then." Micah laughed. Jaiobian nodded, "Yup, you should've been. We were an angry bunch. I was so envious of you Micah. How are your parents?" She shouted. "Uh, my parents are dead, Jai...our home was hit by a drone, and we hid in the cellar. But dad covered us, and, well, mom died a few years later from cancer. It was probably related to the stress. It was a horrible time. Some of our neighbors were killed too. Lots of people were suffering, Jai. But, you've been doing a great job...everyone that I know loves you!" He hugged her once more.

She nodded sadly at his news. She pushed away slightly, "Don't worry, Micah...I have a lot more to do. Are you working? Do you have a job?"

"I have a degree in music, Jai...so, uh...the answer is, uh, no, not really. I play for free mostly, at the hospitals and so forth. Get paid by a friendly smile on occasion...and if the hospital has extra supplies, they'll pay me that way, and, well, I can usually sell those and get enough to live. I'm doing okay, really, better than before, that's for sure."

"Perfect. We need a musical director at Parliament. Here's my card. My secretary will put you in contact with the musical department. I'd love to have you on my team?"

"Parliament has a musical department?"

"Uh, of course it does! Or, it should anyway. Don't worry, let me take care of this at my end...and I'll be seeing you soon." Jaiobian smiled and jumped up in order to give him a hug. He grabbed her waist, and somehow they quite comfortably shared a kiss. Her first. He smiled, she smiled. She ran her fingers over her lips as he put her down. She smiled again. It was the beginning of a new relationship of something more.

Jaiobian began to recognize others in the crowd...many of them had been cruel to her, but she smiled at all of them and acknowledged that she remembered them well. Besides holding no grudges, she was happier than ever before, as if a door had just been opened. An interesting place to be discovered. A place for her to explore.

She had wondered on many occasions how she herself might have treated the Crows had she not of been one. She was currently working for a much larger society, and was adamantly against crime. Perhaps she simply transferred her protective nature over to this now larger group; considering their survivals and the disruptive possibilities of smaller societies such as that of the Crows. She pondered how she of late would relentlessly fight against crime...the very ideals that she had circumvented in order to survive.

She had realized that good and bad are relative to one's perspectives toward benefit and detriment. She also had come to a somewhat conclusion that perspectives was what must be circumvented, and in order to do so, society would be required to bend the rules, create the rules. Make them up as usual, but with a core benefit as would be agreed upon. Everyone in her opinion had worth and could be saved. If she could be, than anyone could be. That was empathy, the ability to be there in another's shoes. To see that they were in fact representative of ourselves. No matter how evil we determined them to be...they were our own probability manifested in human form.

The fitness center was boarded up and a sign was posted that it was soon to be demolished. It would be necessary to dig up most of the town in preparations for the planned eventual bio domes. She sighed and looked back up the street. What an odd feeling it was to be back looking at the dust accumulations of something left to whither.

She wondered her outcome had she remained. Would she had withered? She assumed that she would not have lived past her teenage years as it was highly unusual to do so amongst the gangs in this area. Wars were territorial, and she had been prepared to die if necessary in order to hold on to the specter of reality that the Crows thought they had possessed.

In her mind she envisioned her gang, her family, and her friends being cleaned and prepared for their flight that day. She remembered how dirty and infested they were, with various parasites that found their homes in their hair, clothes and crevices. She shuttered in nearly disbelief how they had all managed to survive up to that point. If not for the doctors at Universe City, they would not have known that Blade had a moderate heart problem, and Craze was a diabetic.

She remembered that there were a few happy moments such as when a sympathetic couple took pity on the Crows and brought several pizzas and sodas. She smiled momentarily. But her mother had found out and was very upset that she hadn't asked for money instead. Her face saddened over the thought of her mother. It could have been different if she had just kept herself clean, but the drugs took her away from even her own daughter. Jaiobian blamed the drugs far more than her mother. They offered escapism into the bliss of its slavery. She had seen her mother perhaps at her worse, but remembered moments of clarity to which she held and cherished those particular memories. Moments when she would tell Jaiobian how much she meant to her. Memories of a mother that did once love her.

CHAPTER 20: THE COLLAPSE

Confusion now hath made his masterpiece.

William Shakespeare

Universe City was not the only reason for world change. The world, our Earth, its societies of species including our own was in torment. The Drone Wars that had unfolded throughout the planet with what seemed to be a panoptical surveillance of diversely equipped camera drones had not only taken its toll on the psyche of human beings, but was successful at generating even more paranoia provoking avalanches of conspiracy theories that many appeared to live by.

Life's' symbiosis seemed a diminishing delusion. Although there was a few surviving species, birds could no longer be heard, nor crickets in many if not most parts of the planet. The occasional trade winds were now wind storms that brought about nearly hurricane gales. Hurricanes increased significantly in intensities and brought almost certain destruction to shore lining societies. Water levels rose covering nearly every seacoast that was not secured by not only breakwaters but embankments that created human developed sand havens.

Many forests had long since been destroyed, and efforts toward restorations were met with overwhelming support by a population in panic. Currency was moving sufficiently as to create a prosperous market. Humans however, continued to move away from its influence as organizations were formed in support of The Unified Coalition's attempts at developing a new form of government that supported altruistic efforts. Wyatt knew that a new human mindset would be necessary, and felt that the development of an actual quantum computer would be imperative toward that objective.

Environmental campaigns had finally succeeded and the Oceans were currently continuously cleaned and salt levels rejuvenated in order to reduce the acidity. Plankton was regenerated using fishponds located in specific organic bio-domes. These efforts however arrived following many marine life species extinctions. Although there were successes toward reestablishing some species, it would take in excess of a hundred years before these attempts truly took on any determined successful accomplishment.

The Unified Coalition by first appearances seemed to be too late, but that was far from being the truth as it was just in time. Just in time to save us from ourselves, but not so much the rest of life. Bringing back species would take time, but that was now available to us.

Artificial Intelligence Operators were numbering in the hundreds of thousands as Congress had originally approved of the first full Manufacturing Plant run almost unilaterally by AIOs. Eventually the Unified Coalition would further develop and increase these manufacturing facilities. AIOs would be incorporated into nearly every human labor intensive activity. There was of course continued concern and doubt toward their capabilities and as to whether they could eventually improvise in regard to possible anomalies. By now most around the world knew of the Caron's worry toward a holonic possibility. It was considered best and essential that humans continue to monitor in managerial positions as we were not only capable of abstract decisions of which AIOs appeared detached, but we would perhaps be the only ones able to recognize such an anomaly as anomaly and act upon it. And although they could perform or imitate convincingly well, we realized that we could not afford any even slight deviations from programs.

With artificial intelligence progressing, the human seemed to become aware that our evolution had perhaps reached its full potential and we were moving toward a quite different adaptation in regard to the Proxies. We were becoming in a sense artificial, merging one might say toward the singularity that had been predicted in which machine passed human intelligence as humans became those very machines. But was this however true? During this period such remained an unknown.

CHAPTER 21: SUDDENLY THEY'RE GONE

No one can confidently say that he will still be living tomorrow.

Euripides

Dominique and Noriko had received their PhDs in Environmental Architectures and were very busy and involved in the construction of Bio-Dome One. A full bio-dome in which the entire population of what was once the east coast of the United States could live within, safely protected from the planet's destructive changes. However, Dominique would fly to what was once California as much as possible in order to be with her husband.

Much of California was now located under the increasing rise of the Ocean. Cities were relocated inward, and plans for a west coast Bio-Dome were in the making. It would take some time before it was begun, as other areas around the world were in more need.

Dominique was on such a visit when their worst nightmare became a reality. "Is Doctor Caron around? Has anyone seen Doctor Caron?!" Screamed a young man that worked in the receiving office at C.I.T.A., as others from the office ran and looked through the rooms. "Have you tried calling his cell phone?" Came a reply from one of Wyatt's assistants. "Of course we have!" "Oh, I see, uh, his wife is visiting, so you'll probably find him at their place." Just then Wyatt and Dominique approached the crowd.

Wyatt was somewhat confused, "What's going on?" Wyatt's best friend Professor Leonardo Star walked up behind him. "Wyatt," Wyatt looked at him with interest, then questioned, "What's this about, Leo? Why are you here?"

"It's about your parents, Wyatt."

"What about my parents? Where are they?"

Leo stood in silence.

"What about my parents, Leo?!" Wyatt's face contorted, his eyes took on concern, his demeanor had changed to nearly panic.

Leo's expression was of compassion as he stood firm. The crowd was silent. "They were involved in a vehicle collision, an accident. They're dead, Wyatt." Leo held his composure.

"What are you talking about, Leo?"

Leo continued, "They were hit head on just about an hour ago. The other driver is also dead, as he was apparently intoxicated and didn't have the auto drive on. They were pronounced dead at the scene.

They're bodies are at the morgue. They want you to come in for the standard identification although most everyone knows who they are."

A loud gasp and chatter came from the crowd, some sobbed, some moaned, but all were visibly shaken over this loss. Some wondered how Professor Star who was just as close to Drake and Vicky as he was to Wyatt, could just come out and say to Wyatt that his parents were dead, as it appeared rather brash, possibly callous.

Dominique stood in shock with her hands over her mouth...mumbling she whispered; "No, they can't be dead, they can't be, Leo...not like that. No. Not like that. Not them. No Leo...we didn't..."

Wyatt stood firm as he looked into Leo's eyes in hope that this was some sort of sick joke. His parents were invisible...they couldn't be dead. Dom let out a muffled scream, as Wyatt collapsed. Leo caught him and held him up by placing his shoulder under Wyatt's left arm.

What seemed like a shock wave, suddenly took its victims as Dominique stood in silence unable to move. Leo called out to her "Dom, we need you...Dom!"

Her instinctive nature kicked in, she blinked and shook her head, and helped by holding up Wyatt's right arm until Professor Ron Caldwell arrived and pushed his way through the crowd in order to help Wyatt into an automated vehicle that was programmed to go to the morgue. Leo quickly entered on the opposite side. Dominique was in thought and moved slowly next to Wyatt in the back seat. Ron got in the front, on the passenger's side. Dom rolled down her window and was given a package of wet wipes by one of Wyatt's students. She thanked her and quickly pulled one out and applied it to Wyatt's face, she spoke, her voice trembling; "We can go home instead my darling. We can do this another time?" Wyatt continued to look toward the floor. "No, I have to know for sure" Wyatt quickly responded.

"Okay. Driver follow program", instructed Leo. The vehicle left the curb and continued onto the underground highway.

Security surrounded the Morgue. Two forensic scientists as well as several other varied physicians met them at the entrance. His parent's bodies were severally damaged and were therefore covered within zipped black body bags. The morgue assistant pulled down the zippers just enough to reveal their faces.

Wyatt rushed to their tables. Standing there for what seemed like several minutes for the others, but in actuality was only a minute, Wyatt looked at the faces of both his parents. Wyatt buckled as his knees hit the hard cemented floor. His face contorted as if every muscle were being stretched to capacity. He looked at his fingers on both hands as if they held some sort of message, but there was none. His hands were as empty as the darkness of space. Sinking his head into his hands, he began to cry. No one until now had ever seen Wyatt cry as he personally felt that crying was self-centered. Now however, he felt the emptiness as it unfolded. He knew that this was it as far as his parent's existences. He would never see them again, nor feel that sense of being the best interest of these people that he loved so very extremely. So, yes, it was indeed self-centered, but they were in fact his greatest supporters, and it was because of them, this self-interest, of what was called love, that they would be most profoundly missed.

It was because of them that he was who he was. It was because of them that he lived to such ideals. And it was because of them, that life had meaning.

Parents, he felt, were the reason that we as humans sought the gods, and held such a gene. His parents were his overseers, which he lived for and hoped to please. How could he go on when his incentives were gone? Why should he? Existential angst filled his mind as he demanded answers. The world in that moment quickly became a cold place with no meaning.

Dominique knew that she had to be strong although she couldn't hold back her tears. She had never seen her husband in so much pain and it seemed to transfer to herself. She wasn't sure what to do. She knelt down next to him. She could feel her body beginning to uncontrollably shake as she looked toward Leo, "Leo?"

Leo knelt down next to her and whispered, "Let's give him some time to say goodbye, Dom. You need to be strong and hold it together for him."

Dominique nodded, and looked toward Wyatt. Her voice trembled, "My darling, you take your time. We'll be in the next room. "Ron turned toward the door, and Dom stood to leave, but Leo appeared to hesitate.

Wyatt gestured for them to return as he slowly picked himself up off the floor by holding onto the sides of the tables that held his parent's bodies. Leo handed him a handkerchief. "Thank you," Wyatt wiped his face. "No, it's okay, you guys should stay. You're family, they love you too." Wyatt again wiped the tears of his face but this time with his hands as he pulled his fingers down over his eyes to his chin. "I'm glad that you're all with me. Wyatt blew his nose on Leo's handkerchief, "No time to even transfer them to an artificial. This is just ridiculous, you know? My dad and mom had so much more to give this world...I just don't get it? It doesn't make any sense. Is this system really that stupid? Just chance and we're just nobodies in a Universe that doesn't give a damn about any of us." How can this happen? I mean, look at them? No time to say goodbye, no time to try to save them....maybe..."

Dominique walked over to him and they held one another for quite some time. She could feel his elevated heartbeat, and the shiver as he moaned. Then, Wyatt began to openly cry, and Dom was a bit relieved as she hoped that he could release more of his pain through tears. "Yes my love, they will always be loved. The whole world cries with us, the whole world." Dominique's whisper was spoken through the thickness of sadness as she bit her bottom lip in order to not break down, attempting desperately to be strong; her eye ducks filled with tears racing over her cheeks and falling to the ground. Wyatt nodded in affirmation.

Wyatt eventually managed to tell the staff to schedule his parents for cryogenics. "We were notified, Sir, that your mother's request was to have her organs donated. However with so much damage."

"Yes, that's mainly why I'd like them to be put into cryo, I know my mother's wishes. Perhaps one day her organs will be of value, but at the moment."

"I understand, Sir. We'll immediately schedule cryogenics for both."

"Thank you." Wyatt put his arm over Dom's arm that was around his waist. Then he pulled away gently in order to bend over and kiss his parents. Putting his cheek on his fathers' he uttered, "Thank you, Daddy, my dad, my father, my Papa, my friend, my mentor, my teacher. You were." Wyatt paused, crying. He cleared his throat and continued, "The very best father that anyone could ever have. Ever Papa! You taught me so much! I'll miss you for the rest of my life! I'll always love you!" He kissed a small portion of his father's cheek that was cleansed of blood by his own tears. Then he moved to his mother, "Thank you, Mommy, my mother, my mom, my mama, my best friend, my mentor. You were the very best mother that anyone could ever have. I love you so much Mama, Oh, my beautiful loving mama! For the rest of my life, I will never forget you, your voice, your smell, your laugh, your smile. I'll always love you, Mama." As Wyatt kissed his mother's forehead (her face bloodied, her nose nearly gone) his tears fell onto her eyelids making it appear as though she too were crying. He patted her cheeks with the handkerchief, and much of it turned red with blood. He folded it and put it in his pocket. "Leo, can I keep this?" He patted his pocket.

Leo nodded, "Yes, of course you can, Wyatt."

Wyatt slowly turned and again put his arm around Dom. Dominique spoke, "They would have wanted us to be strong. They were the strongest most amazing people that I've ever known, and I know that they would want us to go on and continue with their work. We're going to get through this, okay?"

"I know, Dom. They had each other, we have each other. As long as I have you, I can get through anything."

"You'll always have me, my darling. You'll always have me." Dom wiped her tears. "Let's go home. We need to call family and friends before they hear about this from the press."

Wyatt nodded then replied, "Would it be alright if we go to my parent's home?"

Dom nodded, "Of course, Si, I was actually thinking about doing just that, Mi amore. Yes, let's go there." Dominique responded as she looked into his reddened eyes, wiping his face with her fingers.

Ron and Leo left by auto-taxi to their own residences, while Wyatt and Dom were taken to the Caron's estate. Wyatt smelled his mother's vegetable soup simmering in the Crockpot. Tears filled his eyes, as the aroma stimulated his memories. "She made the best soup." His voice quivered.

"She made the best everything, including you." Dom immediately responded.

Wyatt nearly smiled, then nodded slowly as he looked down at the wooden floored entrance. "Yeah, Pops would go on and on about her cooking in hope that she'd prepare something. She always amazed us with her recipes...that's for sure." Memories filled Wyatt as they entered the great room.

"Maybe, maybe we can just pretend like they're here?" Dom asked as she glanced around.

Wyatt responded, "I wish that I could believe stuff like that, Dom, but, I can't. Too much like my old man, I guess."

Dom nodded. "Your mom, she tell me that there's more to this death thing, dahling." Dom bit her lip each time that her accent was revealed, and crinkled her nose. She had practiced English with an American accent vehemently in order to feel more a part of the Carons'. But to the Carons, accents were beautiful, as were languages.

Wyatt reminisced, which caused his facial features to soften, as he slowly spoke, "Yeah, mom was always a little more romantic than Dad, I guess. Dad would usually look over at me and wink, when she'd go on such rants. "

"Your mom had a near death experience, Wyatt."

"The brain is responsible for hallucinations, as it is with the observations of reality, Dom. I could never get myself to believe in any of that stuff, because there's just no proof. If mom shows up, I may change my mind, but other than that, in my opinion, it's a waste of time." Wyatt stepped forward, suddenly a gush of wind blew the two immense entrance doors closed and Dom jolted.

"What was that?" Dom shuttered.

"The tunnels produce wind gusts sometimes when there's a lot of traffic."

"But at this hour, there's not a lot of traffic, darling." Listen?" Dom paused in order that Wyatt should listen to the silence of the highway.

"Hmm, yeah. Ah...I assure you that it was a tunnel wind gust, Dom. Probably an earlier accumulation. I'm gonna spend some time in Dad's laboratory if you don't mind?"

"Okay. Do you mind if I have some of Mom's soup?"

"No, go ahead."

"Are you sure?"

"Yes, of course. Mom would have wanted us to have some. I may join you later."

Dominique walked into Drake's lab a few minutes after visiting the kitchen.

Wyatt looked at her curiously, "I thought that you were going to have some soup?"

"There wasn't any." Dom replied.

CHAPTER 22: LIFE AGAIN

Life is like riding a bicycle. To keep your balance, you must keep moving.

Albert Einstein

With hope in the newly founded world government, and an ongoing change toward the mindsets of the world's human population; we then diverted drone manufacturing toward the construction of AIOs, and we continued and increased the rebuilding of our world.

Doctors Drake and Victoria Caron were credited with a vast amount of innovations. Doctor Drake Caron had not only busily worked on advancements toward the AIOs but chimpanzee proxies.

Following his parents deaths, Doctors Wyatt and Dominique Caron persevered with the Caron's philanthropies. Wyatt along with many in C.I.T.A., continued on with Drake's innovations. And Dominique and Noriko worked relentlessly on the bio domes designs and adjustments thereof.

As if life wasn't complicated enough or busy, Dominique soon announced that she was pregnant and Wyatt for the first time in what seemed like such a long time since his parents passing...felt tremendous elation. No matter their schedules, there was always time for one another and their soon to arrive baby. Dominique had chosen names...if it was a boy, he would be named "Drake Wyatt of course!" after Wyatt's father and Wyatt, and if it was a girl, "Victoria Wyanet" after his mother and his lost sister.

Drake Wyatt was born in the first official east coast Bio-dome hospital. The air was filtered and the foliage was beginning to grow and produce enough oxygen to support the growing population. By the time that Drake was five years of age, the west coast Bio-dome 11 was nearing completion and arrangements were made to move into their new home, which was the original Caron's mansion renovated by Dom, and now set into its own bubble.

The bubble that held the Caron mansion was nearly four acres in circumference with the enormous fifty thousand square foot home in its center. Dominique made sure that their laboratory was located on the top floor with views of their virtual world out its windows. The world within the surrounding bubble was however not all virtual but a combination of actual foliage including what looked like the sky and endless ocean with actual waterfalls, waves, a dock and a beach. The program center was located neatly under the home, and could be accessed in order to update and repair when and if necessary. These bio bubbles along with the main Bio Dome rested on a series of nanotubular coils and layers of nanobotics which could move and thereby absorbed not only sound but any geological movements.

During the building of Bio Dome Eleven, Drake became close friends with a little boy by the name of Benjamin Mitchell that lived near his grandparent's mansion. Ben's mother had recently passed away, and he appeared distanced from his Aunt and cousin. Ben was quite remarkably genius. Most of his curriculum in High School was reverted to the new centralized University. If it were not for his personal circumstances, Ben would have bypassed High School completely. As it was, he and Drake graduated with honors and went on to medical school together, then on to medical research. Medical now included nanotechnologies and therefore research into various methods and applications were required. There was also a series in which students would group and work on innovative applications. This covered what was once called a dissertation in research.

Wyatt and Dominique fell in love with Ben, and attempted to adopt him when he was younger. However, Ben's Aunt refused to allow the adoption. She felt a sense of commitment to her sister. Wyatt and Dom then made sure that Ben along with his family were well taken care of, and a personal dome was built surrounding the home that Ben's mother had left him. Dominique added a few of her special skills to its updates and renovation. Drake and Ben became inseparable, and Wyatt and Dom perhaps not legally but certainly emotionally now had two sons.

CHAPTER 23: THE HOLE NEVER FILLED

The loss of those we love, is as unacceptable as pot holes.

C A Solis

The majority of the human race continued to be in conflict with the few that had chosen to *Burn the World.* Attempting to destroy whomever or whatever they chose to blame, as well as the planet itself, by whatever means possible. Perhaps the anger was understandable as many had lost their loved ones through the numerous battles, and some amongst these, sought, to up rise against anything they might of felt could possibly ease their pain by vengeance. The pain however was never satisfied, as those inflicted by this lust searched out increasing amounts of targets.

The numbers amongst those that fought in the name of myths had declined significantly to where the world became uniformly secular. Much as the holocaust had convinced the many Jewish Atheists; the Drone War or what some called "The Last War" convinced the many throughout the world that the world was what we made of it; deities could not or would not save us from ourselves.

Universe City had proven to be a sensational success. The governments of the world were in the avant-garde of uniting. Ideas were pouring in from all corners of the planet into the new world government that could by reason bring peace.

On the cusp of hope came an unexpected disaster. Water levels around the world were on the rise. The pressure on Universe City's protective walls increased and began to show progressive damage. New leakage discoveries were found daily, and the task of relocation was thought to be too extraordinarily large. Millions of students and hundreds of thousands of teachers, aides, and immediate family, along with administrative personnel and members of The Coalition. Bio Dome One was nearly completed and most of the east coast had been relocated. Moving millions suddenly into this bio dome would prove to be a tremendous task. This is what provoked the new Unified Coalition toward incorporating centralized Universe Cities within each and every Bio-Dome. This was overwhelmingly agreed upon not only by the Unified Coalition but also the people of what was once the United States toward the first of the bio domes...and then eventually it would be voted in favor toward all other Bio Domes.

Dominique and Noriko were the necessary human components' toward the first of the cities. AIOs along with a few million human workers broke ground. Ground breaking had been done not only on the east of the then United States, but in several other locations around the world. The west coast was on the list to be the eleventh of such cities; which would eventually become the Caron's permanent home.

The monetary system was continued in order to allow the human population some form of achievement that could be transferred toward necessities and wants. In order to continue this system of distribution, a supercomputer was put to work that was capable of exoflops or a trillion million trillion calculations a second. However, the Unified Coalition had hope that a faster and more efficient system would soon be discovered once an actuated quantum computer was successful. C.I.T.A. as well as other Physics and Computer departments around the world were working on such projects. Some claiming success only to be dismissed once peer reviews were in place. In the meantime, the Unified Coalition imposed caps on wealth throughout the world, and a more egalitarian system in which basic needs would be met was gaining popularity. The Unified Coalition was quick to establish centralized education access for everyone in the biospheres construction. Universe City would be a featured asset in each and every Bio Dome, in which anyone could freely attend.

The now accepted "Unified Coalition" of the world government was continuing to be filled with seats from Universe City's graduates, and they were tasked with the incredible and complex feat of reestablishing residences for millions of children. Soon Dominique and Noriko were called upon toward further designs. This was after all what they had hoped for, and with the power of not only the Carons but of many members of the UC, they would incorporate their designs toward the first of the planned Domed Cities on the East Coast of the once United States, and then the world.

It was decided that the central area of what was once the United States would be left free to regenerate forests and life. Only the organic bio domes could be seen amongst the vegetation that was attempting to come back. These bio domes would be removed once their success was considered permanent.

Bio Dome One would take several years even with the nearly two hundred thousand AIOs that had been produced and were continuing to be produced. We were learning, sometimes from our mistakes, but we were increasing in efficiency and building on new information. The AIOs appeared to be increasing their own efficiency by self-production, and soon there would be millions around the world working day and night in constructing our homes, and their recharging centers. However, humans appeared to be necessary in order that new information may be generated, and we in a way served these machines by our means of programming their directives.

CHAPTER 24: EDUCATION

Live as if you were to die tomorrow. Learn as if you were to live forever.

Mahatma Gandhi

During the period of the tragic deaths of Wyatt's parents, Drake and Victoria Caron, it appeared as if the world was beyond chaos and collapse was inevitable. The loss of the Carons meant, to many, that… they was no one that could control those that had been produced or and molded by Universe City. These graduates were taking over the world. Procedures were put in place by The COI to evacuate the residents of Universe City into the first of the bio domes, with plans that they would eventually be dispersed to the other bio domes as soon as constructions were completed. But there was some amount of prejudice toward this population.

AIOs were being manufactured as soon as possible, but demand outreached supply. Doctor Drake Jed Caron was now deceased, and his son was in mourning. Congress had overwhelmingly passed the permits toward manufacturing of AIOs, and the COI continued with such permits. But only on the condition that Drake Caron monitor production and insure all security measures were taken. The COI was incapable of meeting this stipulation and the public gave into propaganda. A new leader was necessary, and the world demanded a more unified committee. The COI then became the new "Unified Coalition", and Jaiobian was appointed as its leader.

Wyatt would yet again be put in the spotlight. It appeared that the worlds' population was in panic and a voice, a familiar voice was necessary to bring some sort of order or calm. With no further delay, Wyatt was contacted by the new Unified Coalition leader... Spokesperson Jaiobian Richards. She and Wyatt together would work on bringing the nations toward a stronger bond under The *Unified Coalition*. Wyatt was asked by this new leader, as well as the coalition, to lead. Wyatt denied the request, and this would eventually be the trigger toward a democratic altruistic leadership. He wasn't a politician as much as he was a scientist bent on using science toward clearing the way toward survival of our specie, and what was left of the symbiosis. These needs are what motivated Wyatt to once again fully dedicate his life toward life.

As Universe City was reestablished at Bio Dome One, it became the centralized hubbub. It was the place to meet, eat and be entertained, along with being educated. The newly built Universities that would eventually become the Universe Cities throughout the world developed programs in order to handle the eventual relocations into the Domed Cities.

There were quasi graduations throughout a child's education up until final graduation from University City. However, education never ended. Adults frequently met at the pole for polemics, therefore many poles or meeting places were established in the bio-domes.

Universities also developed special needs programs. This was based on the successful turnout of Universe City, in order that all children develop and meet their fullest potentials.

Universities would consequentially offer a variety of venues of entertainment and business ventures. They soon became the central focus of most cities around the world. If you wanted to see a movie or a play or lecture or spend some time in a library or have a pizza, campus was the centralized location for just about anything.

As ground was broken for the many eventual bio domes, it was decided that each would offer their versions of Universe City. This allowed for cultural diversities, or such incorporations. These universities would be for all people.

The rich oligarchies around the world were actually the driving force toward incorporation of Universe City. With its tremendous success ratio of producing leaders, in particular world leaders, some were motivated by envy in wanting their own children to be educated to such degree.

Our world began to focus on education as war was no longer a factor toward power, as was proven by Universe City. Education was power and with power came great responsibility.

CHAPTER 25: NEXT STEP

The only sure bulwark of continuing liberty is a government strong enough to protect the interests of the people, and a people strong enough and well enough informed to maintain its sovereign control over the government.

Franklin D. Roosevelt

Wyatt would continue to mourn the loss of both his parents, as it made him feel their continued presence. If not for Dominique, he initially could not have imagined going on. Dom kept Wyatt as busy as possible with little time left to feel sorry for himself. Instead she made it obvious that he was more than needed, and needed immediately. It wasn't difficult for Dom to convince Wyatt, as he was left with very little leisure… time, if at all. There was always something that was needed, and now the Caron businesses belonged to him. He was in fact considered the richest man on Earth. However, he constantly emptied accounts in order to finance philanthropic endeavors. The Unified Coalition also capped the Caron's wealth, but still his investments continued to be successful, and it was necessary for Wyatt to steer the monies into streams of benefit toward the planet.

As Bio Dome One progressed toward construction, Dominique and Noriko along with several members of the Unified Coalition coordinated the efforts. It was a monumental task as not only forests but mountains were being surrounded by the enormous nano-tubular dome, with a projected nearly one million underground dwelling domes being incorporated into the design along with connective transport tunnels. All of these particular dwellings would be community based. Neighborhoods were built into the community domes. There were also over thirty million private "bubbles", or domes that surrounded private residences. One of these many personal residents would belong to Wyatt and Dominique. They however would keep his parent's home with the hope that scheduled Bio Dome

Eleven would encompass his parent's residence to which Dominique had started redesigning on the west coast.

Bio Dome One, as with the eventual other domes, was also built for possible space launch. Underground surrounding the dome were hundreds of thousands of space elevators made with nano-tubular construction that would rise up to the occasion. The theory was that these elevators could slowly elevate the Bio Dome into space in which the sun could be used for the initial launch past the planets' threat whatever that could be. The elevators were to be energized by nuclear fusions, or several buried reactors surrounding the domes.

Artificial intelligence operators were in massive self-production, as AIOs built other AIOs to various specifications. Each manufacturing plant had so far diverted design and programming them for sex...as was decided by the Unified Coalition. The UC made the public aware of the possible dangers involved in manipulating such systems.

Despite the warnings, there were several illegal applications that could be purchased through the underground black market, and so far over two dozen deaths had been reported that were unrelated to the aesthetic applications but correlated to the software manipulation, as these artificial intelligences were machines that had highly technical programming; in which any direct manipulations to programming could be recognized as a virus.

What seemed to occur was that the AIO determined the perpetrator as attempting to initiate a virus within its systems and therefore shut down such avenues. However, the human subject attempting to test the AIOs newly initiated programming were met with electric shock enough to stun or kill, as these systems protected its components. Most of these deaths were credited to heart attacks.

This appeared to be a form of survival amongst the AIOs and a new branch of study called "Theory of Holonic Sense" was opened specifically to study and perhaps prepare ourselves toward their possible evolution.

This theory was based on previous studies in which it was considered nearly impossible to program common sense intelligence into an artificial. We could of course program mimicry such that it appeared to be a common sense "response", but not actually common sense. Such that a human would regard as obvious. In example, if a human being existed in a room then the components (our body parts) of that human being also existed in that room. To the human this was a matter of common sense. But to an artificial, it was a matter of tremendous amounts of programming and exaflops' capabilities. To exist was a matter of philosophy and not so obvious to an AIO. Humans accepted and challenged philosophies...AIOs did not. Therefore, if the AIOs displayed what some humans interpreted to be a commonsense response when no such reasoning programming had in fact been initiated...the UC felt that a new field of study was in fact necessary if only to monitor any potential changes if such potential existed.

As more and more AIOs were produced...more and more artificial neurons were being developed, upgraded and used and tested, as these intelligent machines were used in diverse tasks amongst human beings; there were ongoing reports in regard to possible holonic effects. Most of these were assumed to be confabulations, as no one thus far had recorded any events, nor were there any group witnesses.

Financially, C.I.T.A. as well as the Caron Estate and Victoria's Venue (Wyatt's mother's philanthropic organization) were benefitting greatly by the increase demand of not only AIOs, but human's transference into Proxies. Wyatt monitored C.I.T.A.'s progression, and on occasion would contribute what he could to the ongoing projects. It was always appreciated when Wyatt gave his assessments, as his particular genius was so far incomparable. The quickness of the AIOs and Proxies were based on immediate responses to known information. So far, neither could in fact innovate and or produce new information, as appeared was the humans' specialty. Wyatt being extremely proficient at such tasks. The Proxies during their initial transferences were remarkable and excelled over human capabilities…however, that period did not seem to last as they phased more into the mechanism.

With the profits from such project's successes, the reestablishment of Universe City to Bio Dome One was also a success. As planned, Universe City was centralized within the enormous East Coast Bio Dome One, and all humans (children and adults) were welcomed to attend. This however created an unfortunate rush of immigration to the Bio Dome, and even though the Unified Coalition had decided to establish branches of similar Universities as that of Universe City into each and every Bio Dome that was being constructed and or planned to be constructed, the rush continued for quite some time until the others proved this balance to that area. Eventually the population of the world migrated into the Bio Domes that incorporated the aesthetics and sometimes cultures of that region.

Perhaps it was because of the war that many people no longer trusted government. But the Unified Coalition was well aware that another system was necessary to bring back and move forward that trust of the people.

Immigration slowed as civilizations were moved into localized bio domes. Dominique and Noriko worked feverously on each of the first four. And as other environmental architects joined in from every corner of the world, Dominique was able to focus on their future residence in Biodome Eleven. The University of California at Berkeley was established as the Centralized University of Biodome Eleven, and Drake along with Ben continued their educations in hopes of entering this medical school. Because the centralized Universities were fashioned after Universe City, medical schools were far more rigorous and demanding, incorporating advanced technologies and robotics.

CHAPTER 26: IT BECAME REAL

With what a deep devotedness of woe
I wept thy absence - o'er and o'er again
Thinking of thee, still thee, till thought grew pain,
And memory, like a drop that, night and day,
Falls cold and ceaseless, wore my heart away!
Thomas Moore

As Nori and Dom worked vigorously, Dom suddenly took a turn for the worse. With Wyatt, Nori and her boys by her side, she received the diagnosis of AML or Acute Myeloid Leukemia. Everyone was hopeful with the current prognosis. The odds of survival were good, nearly fifty percent if it was caught early into the mutation, and they did catch it early. However, Dom continued to get weaker and weaker with each treatment. On occasion, she would muster up enough strength in order to put the finishing touches on their home at Bio Dome Eleven, as Wyatt worked relentlessly on finding a way to curtail the cancer.

He became obsessed with finding the cure, and such made him look in just about every crevice of the planet. Along with his research, he invited just about everyone with a claim, in hope that there might be something that he had missed.

Soon, Dom would be bedridden in a private room in Bio Dome Eleven's first hospital out of the planned two. Wyatt would not leave her side, as he tried to bring her weight up attempting to feed her at every opportunity, although she was now connected to feeding tubes. She however would continue to diminish, until she was nearly unrecognizable, and refused to eat, demanding that the tubes be removed. As the disease progressed she was highly medicated, her suffering however was unbearable. Wyatt became angry and then angrier, but he could not allow himself to cry, as it would upset Dom and he was fighting for her, even without her being able, and appearing to have given up. Sometimes Wyatt would just talk to her for hours, sharing memories of their incredible lives together, hoping for some sort of anomalous remission to occur.

"If I could only talk to the atoms! I could cure you my love!"

"Talk to atoms?" Whispered Dom.

"Yes, yes, yes, My Love (Wyatt kissed her frail hand)...I only meant that if I could manipulate your system toward stopping the mutation."

"If there's anything after this death, I let you know Mi Amore if you talk to the atoms, if you can talk to the atoms, Si, I tell you." Her voice faded off, as Wyatt continued to gently kiss her hand, recognizing her beautiful accent through her faint and nearly breathless whispers.

95

A little over an hour passed, "Dom?" Dominique opened her eyes slightly. Wyatt could sense her fading away. Her eyes were open, but she hardly seemed there anymore. He picked her hand up to his lips and as he kissed her fingers, they became colder... "My love...please fight...please don't leave me...please don't leave me. You don't understand, Dom, I can't go on without you, you're everything to me, everything!" Wyatt could not hold back his tears. He wept uncontrollably. "Please My love, please, I need you." He noticed their sons through the small glass window of the hospital room, and he attempted to hide the evidence of his pain.

Drake and Ben entered the room. Drake looked with concern at his father. It became obvious to both young men that something was terribly wrong. Drake's eyes opened widely and he moved up to his mother's bed and laid down next to her, putting his arm very gently around her waist, not applying pressure. "I love you so much, Mama", he whispered into her ear. "I love you too." Those were her last words.

The moment had arrived, and Drake cried uncontrollably. Wyatt continued to tell her how much he loved her. Ben stood near the center of the room, staring at first, then nearly collapsing as he moved quickly to the bathroom, holding his mouth he barely made it to the toilet as he vomited. His tears fell like rain into the whirling wind and water of its hydrophobic flush, as his knees felt nothing from the cold tiles of the floor. He felt numb...as though he was in another dimension...muffled...no sense...just there in a place.

No one wanted to move, as if they did, it would then become real.

CHAPTER 27: ONE REAPER DOWN

I'm glad I did it, partly because it was worth it, but mostly because I shall never have to do it again.

Mark Twain

Drake and Ben entered medical school together (both having achieved their Masters in Molecular and Cell Biology). Both with passions for oncology and toward finding the cure toward cancers. Drake chose to be a medical researcher, while Ben appeared to favor practicing medicine, however times had changed and each would enter into the required research curriculum of physics and nanotechnologies toward disease.

Research was now the initial structure of medical practice at the Universities throughout the planet. Both men along with another internist named Amanda Lewis would form a most impressive team.

Their team focused their research firmly toward the cure for cancers. They realized that it would probably entail the manipulation of DNA and RNA in order to stop the mutations, but for now, they steered toward the use of Nano-technological medicinal distribution toward the cancer cells themselves. If they could generate nanobots that could act as iron nano-particles did when they entered cells and destroyed them...they convinced themselves that they could theoretically program their nanobots to enter the human body on such a program of recognition toward the cancer...any cancer mitosis. Theoretically it sounded easy enough, but they needed a primer, a cancer marker...something that would tell them if and when the mutation would begin...and that was a little more difficult, actually a lot more difficult, that was in fact the hard part.

This particular team had somewhat of an advantage as Wyatt and C.I.T.A. were also involved in the research. Wyatt would on occasion visit his son's team and share what information was becoming available.

And then it happened! Their team along with Wyatt found the Rosetta stone of cancer markers! Ectm2 was discovered and isolated by the team. At first it didn't seem impressive or express any known function. They thought it perhaps to be a remnant of past generations. The protein only seemed to express itself in plated and induced cell cultures with rapid growth. Wyatt however decided to look for its expression in cancer cells. With minor adjustments and later that year, Wyatt performed the verification. That was the year, that C.I.T.A. and Bio Dome Eleven's University Medical team announced the *"CURE FOR CANCER!"*

Not only all humans, but without realizing it, all of biology would now be cancer free, as the... medicinal, through the current nanobotic distribution, would... arrive as soon as the mutation began. The programmed nanobots would not decay until they delivered the medicinal and then they would enter the intestines for dispersal. This form of drug delivery was so miniscule that the many nanos that were contained in pill capsules, soon found their way through the molecular structures of walls, and then to the outside world. Once such was realized, the medicinal was manipulated to assuredly benefit other mammals.

The world celebrated!

Drake would continue working on molecular machineries and the miniaturizations of nanobots, while Ben became a full time surgeon. Amanda and Drake were inseparable as she joined him while diversifying her own researches.

The Unified Coalition suggested changing the name of C.I.T.A. to the "Wyatt Caron Institute of Quantum Technology", acronymic as "WCIOQT" and pronounced as "Wyatt." Wyatt was humbled, and of course honored to have such bestowed. The new WCIOQT was designated, and announced to the public.

The team won a few Nobel Prize awards that year, that were bestowed on them not only by the Nobel Committee but by the Unified Coalition's Speaker herself (the highest position on the UC)...Speaker *Jaiobian Richards*. Yes, she had taken Mister Richard's surname in his honor as a child, and he was more than honored that she did so. (He cried as usual when it came to her and as much as he tried to not have favorites, she was his favorite).

As they received their prestigious accolades from Speaker Richards...the Team felt overwhelmed. She was of course, well known, and well loved. She hugged each of them and whispered, "You know who to call if you should ever need anything". She looked at Wyatt and winked, "My dear old friend. You know, I never told you how much I owed to your father and mother. My life. Yes, I owed them my life, and now the world owes you for the many lives that you all have saved. Your parents were beyond great, Wyatt and so are you." she smiled. Wyatt's eyes lit up with pride, "Thank you, I appreciate your kind words", Wyatt replied, then he whispered, "We're not that old." Speaker Richards smiled then cleared her throat, "You're more than welcome, Doctor Caron...I also wish that your incredible wife was here to see this...we at The Unified Coalition were devastated when we had received the news." Jaiobian cupped Wyatt's hands in her own, holding them firmly. "Dominique was most certainly a Caron. I'm sorry to remind you of such a loss, but the Unified Coalition will be meeting in a few days to discuss the arrangements in regard to the statues that are being constructed of her. We'll be contacting you shortly with the dedication details."

Wyatt appeared surprised. "Statues?" "Yes. Four to be precise and one will be placed at the entrance of Bio Dome One, another at Domes Two and Three, and the other at Eleven. Oh, which by the way, reminds me." Jaiobian paused.

"Reminds you?" Wyatt questioned.

"Yes, yes...Doctor Noriko Matayoshi suggested that the Bio Domes be renamed to "Domesticities" in honor of Dominique. We all found the name play to be not only clever, but appropriate, and we've voted its approval. Speaker Richards smiled.

Wyatt appeared to be in thought, then nodded in affirmation, "I know that Dom would be honored. That was very gracious of Nori, I'll remember to thank her...and thank you as well as all members of the UC." Wyatt blinked his eyes to clear the buildup of emotional proof, as one tear snuck past the duck and slowly trickled down his nose. He pulled his handkerchief from his coat pocket wiping the evidence quickly and briefly smiling and nodding once again.

Speaker Richards quickly approached Ben, Drake and Amanda, "I see that you've grown up to be fine people. My goodness, Drake, you're almost identical to your father!" Drake smiled, "That's what everyone says." "I bet they do...and that's a great compliment as your father is quite handsome." She grinned. Drake nodded. "Congratulations, Doctor Caron! However, I must say that it doesn't surprise me. You're a Caron after all." Speaker Richards affirmed as she placed the medal over Drake's head. Drake chuckled, "For looking like... Dad?" Speaker Richards shook her head, "No... For being like your Dad." She smiled...and Drake nodded proudly.

She approached Ben, "I remember you!" she smiled.

He swallowed, "Good or bad?"

"Oh, I remember Dom and Wyatt going on and on about this incredible young man that was so amazing that they wanted to adopt him...and here you are proving that you are such a man! Carons always seem to know a sure bet."

Ben's eyes began to tear. "Thank you... you don't know how much that means to me."

"Oh, I know how much it means to be wanted and loved." Speaker Richards stated, as she shook his hand.

Her focus then went to Amanda, "And... You... of course are Doctor Amanda Lewis?"

"Uh, yes ma'am." Amanda stated nervously.

"Well, Doctor Lewis, we all owe you a debt of gratitude." Speaker Richards placed the medal over Amanda's head then straightened it on her shoulders. "I'm sure that we'll be more than interested in what you and your colleagues come up with next." Speaker Richards smiled taking Amanda's hands into her own reassuringly.

"Thank you so much Speaker Richards. I'm so honored." Amanda's voice continued to vibrate nervously.

What was now WCIOQT would have the full support of The Unified Coalition, as was apparent that they encompassed a relationship toward bringing back prosperity to not only the human being, but the world's life forms. Besides, nearly the entire UC were not only graduates of Universe City but close friends of Wyatt. It was and would prove very beneficial that he had their support on his many and sometimes questionable innovations and explorations.

CHAPTER 28: PHYLLIS

If the atoms never swerve so as to originate some new movement that will snap the bonds of fate, the everlasting sequence of cause and effect—what is the source of the free will possessed by living things throughout the earth?

Titus Lucretius Carus

The world was well aware that several teams of varied engineers and scientists were working on the first actual quantum computer. Many had claimed such success, but so far no one, as of yet, had accomplished such a feat.

Wyatt intuitively felt that the Unified Coalition needed not only their own overseer, but an overseer for the world. A world that had a history of creating mythical overseers', but hopefully would no longer harbor secrets behind such masks. Had an actual overseer existed, the paranoia surrounding Universe City would not have existed, and the attacks against it and varied countries would not have happened, neither the Drone War that had contributed and continued the devastations of our planet.

There were so many possibilities of such a system. Perhaps it could enter our own quantum systems of the brain. Maybe our current computer systems would become obsolete and our brain's ability to produce hallucinations would be used to show us whatever movie we wished to see in a virtual reality that appeared as real as the world in which we lived. Our computers could be as vast as our brain could perform. Yes, the possibilities for such a system could be limited only by our imagination...but in such a reality could it be possible that it was infinite? If so...we were in for something that we had not yet imagined its depths.

Many supercomputers had been engineered...some with incredible memory and speeds, such as the one that Wyatt and his teams used, the Exaflop which could compute a quintillion (one million trillion) calculations per second. These although remarkable, were nowhere near the capabilities of the so far theoretical quantum computer which in its theory could compute in superposition or every probability at the same moment in time.

Instantly knowing the probabilities toward any event was something truly unfathomable, to which Wyatt and Drake worked vigorously with the help of Doctor Trent Keamoku at the University of Hawaii (soon to be Domesticity 5 or "Oh Five" a reversal of "Five Oh" in memory of being the 50th state of the once United States) and recently acquired for WCIOQT, along with his close friend Doctor Shen Lum in China. Trent and Shen had collaborated on a few projects. In theory, such a system would know just about everything before it happened, or at least the most probable. It was almost a sure bet, or one might say that it was the closest thing to a sure bet.

Wyatt had decided that if they were successful toward an entanglement of such a system, that it would be named "Phyllis." Phyllis was the name that Dominique had given to "Physics", Wyatt's mistress according to her accusations. Wyatt found her frustrations entertaining, and would laugh at the thought that she was made from this mistress in a sense.

Wyatt and Drake were busily working on the mechanics of this system, when the call came in from Hawaii. Doctor Keamoku and Doctor Lum had previously received diamonds in which the flaws contained electrons caught in states of entanglements to electron partners at WCIOQT. Such entanglements between pairs of electrons had been achieved many times throughout the years by varied research facilities. However, the theory that Wyatt held was that there existed a field. If such a field was tapped into, we would not only have taken an actual step toward the quantum computer, but we would be opening up to a giant step in our evolution as a species; which appeared to have reached a plateau.

Doctor Keamoku had received a conference call from Doctor Lum and felt that he may be on to something interesting, and wanted to call Wyatt first before saying anything further. Wyatt opened up a large computer screen that was located on the wall of his very generous laboratory upstairs that had been decorated in his favorite colors of emerald green and aquamarine.

"Hi Trent, Shen...what's going on at your ends?" Shen began to speak, "Was just wondering Wyatt, if you and Drake are getting these same results?" "What results?" Wyatt asked. "The quanta seem to be holographical or at least manifesting as such, or on a four dimensional level."

"Ah, yes", Wyatt paused then responded, "we noticed that only today." He appeared in thought, "Wasn't yet sure as the information leaks were going in and out beyond the Exaflops' capability. What's your assessments?"

Shen remarked, "Not sure, but with a quantum computer, I guess we should be prepared for anything."

Wyatt nodded, "Interesting. Drake and I weren't quite sure and I still remain skeptical. We didn't want to mention anything until we had something more."

"Yeah, Trent and I just figured that anything unusual should be brought to your attention, Sir... so I decided on calling you...nothing more observed, Sir. If it had happened again, we would have called you immediately, of course."

"Thank you, I appreciate that. What I suggest at this point is that we keep on recording these events, and see how far the Exaflop can get on the data. Also, I have some people working on a possible microwave interface system that we may be able to hook into this new relay, if that's what it is, well, until it can build a better one. Uh, so far we haven't been able to perfect holograms...getting close but if this new system is relaying holographical images, well we're about to possibly learn a new language. Let's keep the communication screen open, and time the events if they happen again, or do you have it recorded?"

"I have my log right here with me. Trent...did you record yours?"

Trent spoke, "Mine is somewhere on my desk...hold on...I'll get it." Trent picked up a clipboard and checked his watch, "It happened exactly forty eight minutes ago. He looked over toward his atomic wall clock...yup, it's calibrated."

Shen looked up at a wall clock and then down at his log...yup, forty eight would be my calculation."

Wyatt and Drake both looked at their logs...then...looked up at the screen. Drake's voice elevated, "It's entangled! We're in...we're in the field, whoa... how the hell? It's entangled! The holographics are operating as a matter of...uh...choice? Wha...What did you guys do?"

Trent shook his head, "We didn't do anything...you Shen?" Shen shook his head, "No, nothing."

Wyatt nodded, "Well gentlemen, I think that we may have witnessed a birth. We've been trying and trying to get longer durations of entanglements, but couldn't get the particles to cooperate for very long, or long enough. Perhaps because all of us have been doing almost precisely the same task...hmm...somehow... all these particles... somehow clicked it together on some sort of possible foundational level? Hmm...Maybe it was waiting for us? "

Drake began to speak, "So, what you're suggesting, Dad, is that we didn't entangle the system, it did, when the observers, sort of, uh, entangled themselves?"

Wyatt responded quickly, "Yes, yes...if in theory our systems, the human body that is, is somehow entangled to that foundation...then yes...I can see where we could in fact trigger a response, or what we may be assuming to be response is in actuality our observation of our own entanglements of this foundational information field." Shen and Trent nodded. Wyatt suggested, "You two keep on doing exactly what you were doing and we'll do the same here. Let's see if these holographs continue, and exactly what it is that is being displayed."

"Yes Sir, I'm on it." Trent responded. "I'll be doing the same at this end", was Shen's response. Communications were kept open. Almost immediately Trent screamed, "It's happening!" An echo of vibrations was also audible at each location. "What's happening, guys?" Again the echoes could be heard.

Wyatt spoke, "Gentlemen...this may be a result of an entangled field!" Wyatt's voice vibrated. All four men began cheering, the vibrations were now overwhelming as all four men fell to their knees. Wyatt again spoke, and this time very softly." Okay, that was a bad idea. Let's keep the volume down as much as possible until we can figure out how to corral the field...uh...shall we? In the meantime, let's turn off the lasers, and try blocking the diamond's nitrogen centers with your iron caps." The field continued to vibrate with each gesture and syllable.

As Shen, Trent and Drake slowly picked themselves up from the floor, they signaled their affirmations, and a buzzing could be felt from their movements. Shen whispered, "Let's hope that our theories were correct, this is getting scary."

CHAPTER 29: THE MOTHER

Any sufficiently advanced technology is indistinguishable from magic.

Arthur C. Clarke

The Unified Coalition's communication boards were suddenly swamped with calls. Vibration via frequencies or phonons, were not only being felt but heard throughout the planet. By now, it became obvious that whispering helped bring down the echoing noise. Speaker Jaiobian called Wyatt immediately, hoping that he may know what was going on.

Wyatt picked up the phone receiver quickly, whispering to Jaiobian that they might have just somehow triggered a quantum entangled field to which we were possibly entangled.

"Can this be controlled, Wyatt?"

"I think so, Jai...theoretically, yes."

"Theoretically?"

"Yes, theoretically."

"So, you don't know for sure?"

"No, not for sure...but we're about to find out. WCIOQT is delivering the possible interfaces. We've layered silver and silicon to hopefully distribute the field in a uniformed manner. If we're correct...these vibrations will tunnel into the interfaces. It's gets a little complicated after that."

The Speaker continued to whisper, "Sheesh Wyatt...it then gets complicated? Well Wyatt, I trust you more than anyone I know...but make this work, please, please make this work. I can't bring a theoretic to the people. I have to be able to explain at least a little of this...uh...phenomenon."

Wyatt whispered back, "I understand Jai. You won't need a call from me if it works...everyone should immediately know."

"Good luck, Wyatt. The UC is, uh, quietly on hold at the moment."

Juan opened the Caron's mansion door in order to allow the delivery of the interfaces to Wyatt's lab upstairs. They were small rectangular objects holding thousands of layers of silicon and silver, like corridors which theoretically would allow the energetic vibrations to enter and move in a precise path, to which hopefully these physicists could then determine information, although such was theoretically infinite. They had theorized that electrons would form a two dimensional liquidity as in a superconductive state. If such held true then they truly did enter another dimensional avenue toward not only information but energy. One of the two scientists that was holding the interfaces from WCIOQT ran up the wide stairway to the laboratory and knocked on the door. The vibrations from his actions nearly made Wyatt and Drake lose their balance, each grabbing hold of their desks.

Drake quickly but softly walked to the door and whispered, "You're welcome to come in, but be very quiet please, we haven't been successful at capping the energy as of yet."

"Oh... yes Sir, shhh, I understand...so sorry." The young scientist whispered putting his finger to his lips and appearing embarrassed.

Wyatt nodded and took the interfaces from him as he neared his area. With no hesitation he hooked place one into the system as Drake, Trent and Shen along with the young scientist from WCIOQT appeared to hold their breaths. Suddenly the vibrations ended. Wyatt then place the rest of the interfaces alongside the first in hope that all would be entangled.

Drake whispered, "Did it work, Dad?"

Wyatt whispered back, "I think so." He repeated it louder with his own reverberation, "I IIIII...thiiink... sooo!" then smiled.

Everyone except for Wyatt, cheered. Wyatt took in a deep breath and exhaled, relieved.

Wyatt looked at his computer screen, "Trent, Shen...get ready to remove the caps on your ends...theses interfaces should be sufficient." Each was using an AIO to perform tasks near the energy sources. The AIOs were instructed to remove the iron caps that were theoretically holding the foundational quanta qubits in the diamond's vats or as some would say...the diamond's flaws.

"Okay, the caps are removed." Shen announced.

"These too." Trent affirmed.

Wyatt looked at the interfaces intensely, "Okay gentlemen, let's see how much this mother knows. Interfaces now engaged. Hello?" The androgynous selected voice of the interfaces immediately responded, "Hello."

Trent had his hands excitedly over his mouth. "CRAP!" Wyatt looked at him on the screen, "Sorry, Sir."

Shen spoke in Mandarin toward the one interface that Wyatt was now holding, and the interface responded likewise.

Wyatt asked a question in Latin, "Quid est nomen tuum?" (What are you called?)

"You will decide to call me "The Mother", responded the interface.

"You said "me", do you identify with a self?" Wyatt questioned.

"I am the self-identified as Wyatt Drake Caron."

Drake on a hunch interrupted, "Who else?"

"I am the self-identified as Drake Wyatt Caron."

Trent intervened, "and who else?"

"I am the self-identified as Trent Ikaika Keamoku. I am the self-identified as Shen Inoi Lum. I am the self-identified as... "

Wyatt nodded then interrupted. "Okay, that will be enough...we not only have entanglement...we're all apparently... entangled, it could go on indefinitely until it reaches its last entangled point."

Drake asked, "I wonder if there's a limit?"

Wyatt rubbing his chin, spoke, "Hmm...Theoretically...nope. That's unfathomable perhaps, except that, it's now a reality."

"Dad...that means that it's also interconnected to all life, like everywhere, uh, the Universe? But why then is it identifying as each of us? And not as a unified entangled system of say...one Universe?"

"Each of us act as an interface for particular language exchanges. Points of information, one might interpret this to mean that our brain is talking back to itself as a means of transceiver, one of perhaps infinite transceivers that are now interconnected through entanglements. Heck, the word interface doesn't hold much meaning anymore."

"So, right now a grasshopper could be talking to itself?"

"Yup, but only if it had a means to control language feedback, such as we have done with the mechanical interface. Uh, I'm guessing, as I'm not currently sure that we're controlling anything as of yet. "Wyatt again rubbed his chin as he thought.

Shen interrupted, "Wyatt, do you suppose that it might have a mind of its own?"

"That's the winning answer, isn't it, Shen? I suppose that if there's even two working minds, then one could assume, that, as being a separate field of thought from a singular mind. So, if this functions as an entanglement of all minds, than certainly it could not only appear, but be a separate form of intelligence, as none of us currently can use every mind."

Shen nodded in agreement, "That's one hell of an intelligence, Wyatt. We might have triggered a form of Panpsychism?"

Wyatt replied, "Uh huh, yes, or it was there the whole time, beyond anything imaginable, I'd dare to say. But I can imagine no more wars, no more diseases, no more death...maybe."

Trent interrupted, "We'll, I like that! Yes indeed! Now where do we begin?" He smiled as he rubbed his hands together in contemplation.

Drake spoke, "Hey, doesn't it seem a bit ironic that we once created false gods, and now we may have created a real one?"

Shen's expression took on concern, "Yeah...ironic."

CHAPTER 30: MEETING

Walking with a friend in the dark is better than walking alone in the light.

Helen Keller

Wyatt's personal cell phone rang, and he recognized the number immediately. It was his father's and his best friend and confidant from his teaching days at Berkeley, Leonardo Cameron Starr. "Leo! You couldn't have called at a worse time. I'm currently working on a project that I, uh, I must continue on, and you might say that I can't get away from." (Drake chuckled as Shen and Trent shook their heads at the attempted pun and humor.)

"Perfect! Would you mind if I joined you, Wyatt?"

Wyatt fully trusted Leo, and felt a need to have him there with him, although this project was of utmost privacy. "Yes, Leo, by all means, please do join us. Do you still have my parent's keys to this place?"

"I sure do. So, you've never changed the locks?"

"Never needed to, this place is always open now that it's in a Biodome. Only one way in and that's protected by AIO security. But your keys will be a nice addition to my parent's collection here. I'll let Juan know to notify the other AIOs."

"You and your parents will always be more than family to me, Wyatt. I was overwhelmed when they made these keys for me. Don't know if I can give them up."

Wyatt laughed, "Okay, then keep them."

"No, no, you can have them."

"Keep them, I'm serious."

"Nope, you want them, you can have them, and besides, I'd give you my heart if you needed it.".

"Leo-ooh, you're doing it again...I swear, you're the only one that can do that to me."

"What's that, Wyatt?"

"Drive me bat shit crazy!"

"Do I have those keys too?" Leo grinned as he sarcastically jested.

Drake, Trent and Shen were now laughing as Wyatt shook his head, laughing as well. After all, he loved Leo, not only as a friend, but as a brother. He then ended his conversation, convincing Leo to join him as soon as possible.

Everyone that knew Leo knew that he carried the same sort of dry humor as Wyatt. That was perhaps why the two hit it off immediately after meeting at University. He eventually became a best friend to both the senior Drake and Wyatt. He would on occasion sit in on their researches, in which his advice in particular if any, was always welcomed as it had always appeared beneficial on a quite remarkable level. If not for Leo, Drake would not have been able to find some of the problems with the chimp proxies. Or, how to further enhance the information exchanges of the AIOs. He became a father figure to both Drake and Wyatt. Wyatt graduated with his Ph.D. at the age of fifteen, and began teaching soon thereafter. During Wyatt's studies, Leo became his tutor, and it was beneficial that Leo also taught at the same University that Wyatt would enter as Professor.

Wyatt was curious as to what Leo may say on seeing these results. Wyatt was always welcoming to opinions, no matter who from, he viewed people as some people viewed books...they were stories of information to be read and possibly understood. Leo however, was more of a mentor, which Wyatt's genius seemed to need at moments. In particular to periods of frustration, in which there appeared to be no avenue other than one that offered a dead end.

The transit system was nearly completed, and Leo arrived in less than ten minutes via AIO driven commercial vehicle. The UC was gearing toward a new bubble design for all vehicles that would move rapidly on electromagnetic pulse. So far, only commercial bubbles were available, but soon each individual would harbor their own.

Leo exited while thanking the vehicle.

Wyatt and then Drake welcomed Leo with a handshake and hug. Wyatt pulled back, "So, do you think that the vehicle may incorporate the holonic effect?"

"Perhaps." Leo grinned. "By the way dearest brother, friend, son...you do realize that a family biographer should not be away from his subjects?"

Drake responded, "Yeah, so why the absence?"

"Ah, just needed a little time to slip away from this reality for a while."

Wyatt and Drake nodded. Wyatt responded, "Oh, we all know that feeling, I'm sure. Glad to have you back with the crew." Wyatt again hugged Leo patting his back in reassurance.

Leo looked over toward the laser mechanisms, "So, tell me, what's going on? I figured something was up when I felt the change in atmosphere, and I immediately thought what the hell is Wyatt up to now?!" Everyone including Leo laughed.

Wyatt responded, "So did the UC. Well, we can laugh about it now, but we really weren't sure as to whether we could control the results."

Leo grinned, "You did it, didn't you?! An actual quantum computer"?

Drake responded, "How did you know, Uncle Leo? It's like you have some sort of ESP when it comes to Dad."

Leo's eyebrows rose slightly, "Well it's not so difficult when you know how hard your grandfather and pops here worked on the theoretic. Your grandpapa got pretty darn close with the AIOs...they're about as fast as a classical computer system can be, even with a hint of quantum technology, but he wanted them to be able to problem solve through some sort of emotional overlay, like our abstract brains can. They even messed around with neuromorphic designs, but couldn't quite manufacture neurons the way the brain does. But I'm guessing that this can of worms will be able to do just about anything?"

Wyatt responded, "Here's hoping that it can open up the Universe to all of us."

"You never did think small, Wyatt." Leo grinned. "Well, let's see it turned on, shall we?"

"We're actually not sure if it's turned off, Leo. I can stimulate the diamonds, maybe it's sleeping, if it does such. It definitely can however communicate in every language, and it seems to talk back to us as ourselves, through the interface systems."

Leo appeared to be in thought, "Time to put it to the test."

Wyatt responded, "Just to let you know...it told me that I would be calling it "The Mother." I know that it was correct the moment it said such...how I know...I'm not sure...but I know. I'll suggest that to the UC as it sounds more welcoming than Phyllis...and besides, I know that they'll agree, as it's the name that we'll be giving it anyway." Everyone laughed. "It theoretically doesn't control time, but it knows probabilities, infinite probabilities...so it may be able to intervene. Hmm, that's something that we'll have to bring up to the UC." Leo looked at Wyatt and nodded. Wyatt announced, "Okay everyone, here it goes!"

Wyatt, Drake, Trent and Shen opened up their systems, but nothing seemed to happen. Wyatt spoke to the interface, "Are you there?"

"Where would that be?" came the response.

"I don't know, you...tell...me? Where are you?"

"I'm here." was the response.

"Where is that?"

"Where you are, I am".

"I would like for you to answer some of my questions?"

"Yes. Yes. No, not for another 23 years. Spain. In five days. Yes. ...". The Mother continued.

"Please stop...and thank you for your responses. I'll have to write down my questions later."

Leo spoke, "Wyatt, I think it's a good time to contact the UC and let them know exactly what is going on."

Wyatt nodded and took out his cell phone from his pocket dialing Jaiobian directly.

"Hello Wyatt, you said that I wouldn't hear from you if it worked, and it seems to have worked?"

"Well, Speaker Richards, you may want to call the UC together, as we now have a quantum computer in... need, perhaps, of... some guidelines."

"Oh my! That's, that's, incred ... I'll call them immediately and let you know when we all can meet. Thank you for letting us know so quickly. Is it as amazing as was predicted?"

"Uh, let's just say, you'll probably all want to meet it for yourselves, but it already knows you, really knows you."

Jaiobian's voice trembled with excitement, "Oh gosh...I hope this is it, Wyatt! Can you imagine? No secrets! Oh my goodness! Okay...let me get a hold of everyone...I'll call your home line so we can conference as soon as possible!"

The team went silent as each member thought of the implications.

CHAPTER 31: CONFERRING

Never limit your challenges, instead challenge your limits.

Unknown

It took nearly an hour for Jaiobian to locate all the members as some were on vacation. A rarity for the members, but Jaiobian insisted that they take a little time for themselves whenever possible. The world was indeed under turmoil, but she needed everyone to perform at their best, and sometimes that required rest.

The United Coalitions' Representatives began to sign into the conference, they were greeted by Wyatt's team. Wyatt's screen began to fill with faces. Five members represented each Domesticity. The Domesticities that were not yet built, were represented by area as countries no longer existed. Where there would be required more than one Domesticity, there was assigned five Representatives to each.

"Hey Dad". Suddenly the screen became a holographic projection and the Representatives faces moved outside of the screen. Wyatt looked at Drake. "I swear Dad I was just thinking wouldn't it be nice if we had holographs. How the hell?"

"We're going to have to be careful with the current entanglement to ourselves. I'm not sure how many can affect The Mother, but that's something that should perhaps be controlled if possible. It apparently can scan our brain waves to the extent of imaging our thoughts possibly before we ourselves are aware, that we're aware of the thought. That may certainly be a privacy issue." Wyatt approached the rolled out screen to examine the projection coming from the wall. "Hmm, interesting. Our screens can project three dimensionally but not like this...it's projecting everyone as holograms...remarkable."

Jaiobian was on line, "Yes Wyatt, we probably should put some controls on the QC immediately. Representative Caribbean Raban, suggested that we should begin by designating personal interfaces to which the QC could entangle with whomever possessed an interface. I agree with her, and think that's a good idea. At least that way it would be somewhat regulated. That's if the system cooperates. We have so much to learn. We should be cautious."

All members of the UC were currently on line and several were talking at once. Most were confused.

Jaiobian stood and spoke as she pounded the gavel on the bench, "Please everyone, can I please have your attention, please...your attention please." She paused until the chatter quickly quieted. "Now that we have everyone on line, Doctor Caron and his team would like to make an announcement. The floor is yours Doctor Caron."

Wyatt stood next to the system, "What we now have here, everyone... is the first and only actual Quantum Computer." Many of the Reps began talking at once, however Jaiobian had turned off their abilities to speak to one another resulting in some control over the meeting.

Representative Carlos Santos pressed his button to speak as had many others. Jaiobian spoke, "Uh, Carlos, I mean Representative Santos, you now have the floor." She switched the system over to Carlos.

"Uh, thank you, Speaker Richards. Doctor Caron what limits can we expect of this computer? There's been several claims towards actual quantum computers, how is this different, other than what is appearing obvious?"

"We currently don't know if there are any limits, Representative Santos. We're sure of the entanglement, but to what extent, we're not sure. It may be entangled to everything, everywhere, but it's selective, or we may be directing it. It may be unreliable as one might say that the electrons can mess with its qubits components and make them just as unpredictable as neurons. It's quantum after all, and it may take quite some time before we know how to benefit from it, or it could or might very well destroy us if used against us. We do know that this is the real deal. What we didn't know or weren't sure of before is how the entanglements would affect our own particles. We're still not sure how deep this rabbit hole goes, you might say."

Carlos spoke, "I see. It would be wise of us to put restraints on this immediately?"

Wyatt responded, "Yes, I agree. If we can."

Carlos spoke, "Doctor Caron, any suggestions on how we would go about doing that?"

Wyatt in thought nodded slightly, "hmm, we can ask it how, as it knows the answers. And, that's before we presumably ask them."

Jaiobian interrupted, "Would you mind doing that at this moment, Doctor Caron?"

"I should perhaps mention that it told me that we would choose to name it "The Mother", announced Wyatt.

Jaiobian reacted, "Oh? Hmm, I like that...yes...I do like that... the population may be more receptive to it if it sounds encompassing and loving or at the least, benign."

The Reps of the UC all called out their agreements. "Let the record show that, The Mother it is", announced Speaker Richards.

"Doctor Caron" Jaiobian called for his attention.

"Yes, Speaker?"

"What if, we decide to choose against what The Mother has stated what we will choose?."

"I'd guess that it would just recalculate the probabilities based on our change. But it's tapped into our current probabilities, so changing our minds wouldn't benefit us as the information is what we hope to achieve. That is, in regard to knowing information immediately with no delay factor, it would be in our best interest to perhaps accept these probabilities, less we may find ourselves playing a game of chance once again. If I'm correct, The Mother can give us the most likely probability...the pattern that we as a species are commonly in search of."

"I see, thank you. Wouldn't that however be deterministic? Making us somewhat puppets destined to live out such probabilities?" Jaiobian appeared concerned.

"Nah, not really. They're our choices... and we still have to deal with uncertainty quantum mechanically. You might say that The Mother through time looping will see percentages of infinite possibilities, and based on everything that it is entangled to, it should be able to give us its best guess."

"Guess? Do you mean the most likely probability?"

"Yes, exactly Jai, I mean Speaker. However, it will be so extremely fast and accurate that we may feel like it can't be wrong. And chances are that it won't be. It's almost a sure bet."

"Would you mind now showing us how this works?"

"Of course, Speaker. We're not really sure ourselves, but I'll talk to it and we can see what happens."

"Mother? The Mother?" Wyatt asked. The current large computer interface reacted by vibrating.

Wyatt formed a request, "Please speak."

"I am capable of any language, you preferred English with female tone." The Mother's voice appeared to come through the interfaces.

"Yes, English is fine, and thank you for sounding female. We are hoping that the population will accept your entanglements."

"I will speak in each individual's language as you will wish shortly. This will make the Unified Coalition's members relax. They are all highly stressed at this moment."

"Alright, do so, please. Yes, it's natural for the human being to be stressed when we don't understand something."

"Yes, the basis of survival is such foundation. I am speaking that which you will request." This time the voice came through the individual computers of the Representatives and each Representative heard The Mother explain that it was entangled to them in their own cultural language."

"Thank you." Wyatt although a polyglot, was somewhat dumbfounded as these languages were being spoken by The Mother all at once. He however was able to understand some of what was being explained.

The Mother went on to answer each Representatives questions before they said anything. This was frustrating to Wyatt as he hoped to hear all the questions and their answers.

The Mother spoke to Wyatt through his interface, "Doctor Caron, do you wish to have a transcript?"

"Yes, Mother, I would very much like a transcript."

"One will be provided to you shortly through your computer."

"Thank you, Mother. As you know, I was going to ask if such were possible."

"Yes. I know."

As the conference came to its end, The Mother was fully aware of the probabilities of the entire UC, and it started to initiate a set of principles and limits to the specifications of the UC.

No one knew how or why The Mother would cooperate, but they were all certainly glad that it did.

CHAPTER 32: THE WORLD CHANGES

In a chronically leaking boat, energy devoted to changing vessels is more productive than energy devoted to patching leaks.

Warren Buffett

The world population was by now aware that something was amiss. The recent display of vibrations throughout the planet was enough to get everyone's attention. The UC therefore called a symposium in which issues would be brought to the vote.

As soon as the announcement was made that a quantum computer had been created...chaos erupted as conspiracies ran amuck. Paranoia was now taking on a more menacing face with the general public, and now appeared to be out of any sort of control; possibly because many had no trust toward governments to which was in the slow process of recovery through the established Unified Coalition.

The UC had several discussions on whether a quantum computer was a good idea, would it in fact be beneficial? Humans appeared to like having secrets, perhaps as a personal and individual privilege in which could be hidden insecurities and indiscretions.

Of course The Mother was not absolute, because the Universe was not absolute. The Mother could only discern amongst the most probable amongst established patterns of information. It was however the best guesser that we had, as no other computer was capable of guessing based solely on probabilities. It could in fact be wrong, but would be correct on most so called "guesses." Its best guess ratio was holding at 97% to 98%, and only because some of us were deviating from such objectives as best we could. One could therefore say that we could bet on it to be correct. Other than that, we could not fully base our judgments on this system, we had to keep our minds open to possible and sometimes what appeared to be impossible anomalies.

Due to the overwhelming negative responses, the WCIOQT team was asked to turn it off until further notice. The problem was; no one knew how to turn it off. Wyatt decided to take apart the system, increasing the temperature, attempting to misalign the atoms that the Bose Einstein Condensate was aligning. Although every aspect of the system was brought down to basic components and disrupted, The Mother continued, as the atoms continued to align. This was beyond scary for many. Our best semi-conductors still needed extremely low temperatures, but The Mother apparently did not…it could operate in relatively any temperature…perhaps as long it was entangled to that system. Wyatt however sought to find out more. The Mother endlessly intrigued his intellect and he worked throughout the nights attempting to figure out exactly what was happening.

The Mother could communicate through any technology be it a cell phone , landline, or computer interface Throughout the many systems around the world, The Mother connected and began to manifest as various holographic displays, which were quite incredibly realistic. Wyatt's team theorized that it was continuing its process of entanglements, and perhaps it might be necessary to wait it out. But it was frightening to humans especially when it entered our minds. Most became discombobulated, as though everything were hallucination. Would it reach a point of end? Wyatt and his teams worked vigorously to find out how to get it under control of some sort.

The Mother as it would turn out, appeared to be here to stay, we would decide it best to incorporate it into our lives. This was apparently an intelligence that we at this time were perhaps quite unprepared for, a purely logical mechanism that could reason and by all appearances seemed to be aware, self-aware. How many selves was an unknown. The Unified Coalition decided that it would be best if only a few, those specifically with an interface, could in fact "request" of The Mother. The Mother would be off limits to anyone else. One would have to bring their request directly to The UC before any interface to The Mother would be allowed. The Mother was then requested by the Unified Coalition to abide by this request, and The Mother agreed. The Mother was also requested not to attempt to divert a human's free will. What surprised many was that The Mother was in fact aware that humans did indeed have such a trait. Only the Unified Coalition and only by majority vote could the requests to The Mother be circumvented, and only in best interest most beneficial to the majority of humans on the planet.

These limits were felt to be necessary toward human prosperity, and once they were incorporated into the quantum computer, the world population began to accept the changes that The Mother brought forth.

Wyatt remembered how his father Drake Caron had once attempted to save a chimpanzee's life by incorporating some of another chimp's brain. It was considered a successful transplantation, but the surviving chimp only lived for a few days. If Wyatt was correct, in that two entangled human brains would operate as a separate form of intelligence, than it was unfathomable how the possibility of an entangled Universe would debut. The Mother was to Wyatt the greatest mystery of the Universe, and he wanted to know all the answers. He thought to himself, "Oh how Dad would have loved this!" and he grinned in satisfaction.

CHAPTER 33: NO SECRETS

If you want to keep a secret, you must also hide it from yourself.

George Orwell

Surprisingly to the representatives of the UC, the changes were beginning to be accepted by the majority of the population. Almost immediately the veils of secrecy were dispended. Humans became well aware that they we were being watched and discerned as personal reports could be seen by individuals through their classical computers. However, we were satisfied that we held some amount of privacy from one another. No one person knew everything, as The Mother did.

One of the outcomes was that humans began to disregard public mannerisms and instead appeared to treat others more as they would family.

The Mother would not take sides, but it did function on what was logically beneficial toward survival of this specie, as well as other species. If one needed additional points, one only needed to provide benefit in some sense that was conceivably understood as such by our specie. The Mother was aware of all information available in regard to our laws and ethics.

Law enforcement would eventually arrive before crimes were committed...and warnings would be issued. The War ended and the military disbanded as The Mother informed all sides of probable outcomes. Even personal relationships were subject to probabilities of which The Mother began to be depended on. Some of course decided against the odds as such was considered a gamble even with the best of odds.

All previous monetary amounts were at this time transferred into points, and all paper monies were considered useless. Gold was no longer what determined wealth, but was needed for some technologies and medicinal deliveries. There was, however, an abundance… that had been discovered on several asteroids. We were quick to send a few of our, "first off the assembly line AIOs" programmed in extraction techniques. Once considered rare elements, were arriving by delivery spacecraft nearly daily.

Before the Unified Coalition had gathered together for another symposium in order to discuss what sort of governmental structure was necessary now that The Mother existed; The Mother had organized what would be called an Altruisticity or a government based on the beneficial acts for others, and it was up and running. The more one did for others, the more one would gain points. Points would be necessary toward wants…however, all basic needs would be supplied by government.

The Mother would be responsible toward the recording and distribution of points according to beneficial probabilities and or actualities.

The Mother also informed the Unified Coalition that it would cooperate and not interfere with human's free will, as that apparently was what made the human…human. Other than free will, the human appeared mechanistic. Humans, by majority vote, made it clear that free will would never be infringed upon, other than criminal acts. The Mother, however, could… also disarm and prevent detriment if the subconscious of the human was in agreement with the conscious.

Why The Mother cooperated was a conundrum to perhaps everyone, but we were grateful that it did, and nearly everyone wondered how far it would take us, once we were all on board.

The Mother through the AIOs, began the work on assigned interfaces that would establish the direct connection to The Mother. Following these constructions, The Unified Coalition instructed The Mother to only communicate through these interfaces. It would be decided by the UC whom exactly would possess such interfaces. The operators of these interfaces could then request information and help directly from The Mother. It wasn't yet known how much The Mother was capable. Or as to whether these interfaces were in fact necessary.

Once information became available that the quantum computer would have interfaces, the demand to possess one was overwhelming. The Unified Coalition was forced to make it very clear that whomever received a majority vote from the UC would be such recipient. In order to receive such an audience approval, one would be required to prove the merit of their projects. The Mother of course would be used in the decision as to whether such projects were in fact probable.

Wyatt and Drake were amongst those that would be given permanent interfaces to aide with their researches. The interfaces would thereby become a most prestigious award. The public's curiosity rose each time an interface was awarded, even though such interfaces did not as of yet exist, all in await for the announcement into eventual varied discoveries.

It had taken a little getting used to the point system that The Mother was distributing and recording. Some humans understandably became agitated when points would be removed. Reports were continuous through classical computer systems and the numbers were staggering, thus The Mother was requested by the UC to reduce the reports to an understandable language. We could then request our report at any time, and thereby know how many points we had accumulated to buy whatever we wanted.

Controversial was the decision to make all medicinal including what had been illegal, now legal and available. This was decided mostly because any human could use the new regenerating machines and 3d printing machines to produce nearly anything that was necessary or of want. These machines were capable of locating any program that was requested, and there were so many under diverse interpretations, that it was nearly impossible to control or put into law any controls on any substance. It was therefore thought necessary to regulate some form of control by initiating clinics in which at the very least safety measures were in place. These hallucinogenic facilities called "Highbeams" became very popular, in particular when The Mother was incorporated in order to control detrimental possibilities.

The Highbeams were a crapshoot which humans sometimes appeared to prefer. Before entering such clinics, one signed a release form that allowed The Mother to intervene if the hallucinations continued, thereby diminishing the chance of permanent schizophrenia (which had also been cured through The Mother's entanglements).

Virtual worlds were very popular as anyone could take a trip to anywhere with the aid of The Mother. Our brains could actually connect into programmed worlds. Unfortunately, some humans would remain in these worlds believing them to be reality and there was little that anyone could do, as they could not be removed else their brain would be confused enough to shut the body down, and in most cases, enter death. Although such was of course considered detriment, The Mother did not help with this because the person's subconscious and conscious minds did not request or allow removal. In fact, both usually held fast to the illusions.

It was believed that eventually the mind would tire of the virtual worlds and begin to question the reality. This did happen on a few occasions and The Mother was successful in removing these few from sometimes years in such scenarios.

AIOs used much larger replicator machines for the inner buildings under the domes. What once took months was now taking days and sometimes hours. AIOs major labor intensive usage was on the actual dome itself. Several months and sometimes years were required in order to integrate the translucent graphene and interwoven nanotubes into the enormous structures that entered deeply into the planet.

With the government taking care of basic needs, many humans ventured into the arts, with many becoming creative in an array of ways. As long as one contributed to others, one earned points...even if it was to only open a door, or offer a helping hand.

Many more entered the sciences, in hope to see or discover what no other human had so far ventured to find. Many humans took to the addiction of possible discoveries which seemed to be endless. Laboratories were supplied not only by WCIOQT but by all centralized Universities in each of the Domesticities. It was therefore relatively easy to obtain a research facility. There were also research facilities outside of the Domesticities. Scientists were permitted to remain past the two weeks maximum of citizens requesting outside visits. Eventually fewer would request such visits, as Domesticities provided whatever one chose as beauty, in not only the community domes but individual personal domes.

Yes, the AIOs were indeed required for hard labor and several hundreds of thousands were working mostly on the construction of Domesticities. Humans, however, were also required in supervisory and management or what was now called "director" positions. The Mother kept records of progression, and these humans earned points accordingly.

It was easy enough to earn, but also easy to lose points depending on what we generated toward benefit of our species and the world in which we lived. This kept humans from taking on roles of masters over the AIOs. AIOs were not created to be slaves, but to be support systems.

If one had the original currency of their former country, it was transferred over to the point system, and was subject to gain and loss depending on behaviors toward others.

Teachers, Scientists and Artists of all sorts, consistently earned the most.

Eventually it appeared that everyone was adapting to the new system, and for the most part, the world thrived, our race survived and there was peace.

The Mother increased the AIOs efficiency drastically as they began to appear more and more human by the perfection of the organic field of artificial skin. They also seemed to be expressing the intelligence levels of nearly The Mother, yet not quite the qualities. Their original programming was classical, so this was remarkable although confusing to some, as to be such a possibility much less a probability. This also introduced a new concern by humans, and therefore The Mother brought forward the new neck implants that would be requested in order that humans would have competitive advantage given our abstract qualities. These implants would allow The Mother to educate children and adults up to the University levels. They were implanted only by requests of parents, or in the matter of adults, they requested the implants for themselves.

Universities which branched out from Universe City were the centralized hubbub at each and every Domesticity around the world. It was decided by the people and the UC that the Universities would be utilized for education to all, as humans needed one another in order to truly flourish.

The ongoing construction of individual Bio Dome bubbles underground and within the Domesticities meant growing populations that would need the Universities as a means of not only education but of human integration. As humans became more likened to machines as Proxies, we quickly realized our need for one another.

Proxies were also becoming more reliable with the help of The Mother in our transferences over to artificial. Although there always seemed to be some sort of variation in personalities and a momentary collapse into dementia which appeared to last as much as two weeks. The brain did however appear to transfer most of its information. However, there always seemed to be a percentage of confusion from the original human, and that percentage to the human at this point in history was more than necessary and highly detrimental in determining if the personality had indeed transferred.

Wyatt worked vigorously is his attempts at curtailing death. His passion was perhaps somewhat based on his personal experiences with the Reaper, and he was determined to kill it, kill death. Soon, the transferences over to artificial proxies would be considered a success by the UC, as most of the original personality did seem to take hold, enough anyway for the UC to determine transfers and Proxies a success.

The world celebrated the "Conquering of Death", and the Carons along with WCIOQT was given full credit and appreciation.

Many detrimental conditions that once plagued humans and animals were almost non-existent with the help of the clean atmospheres of the new Domesticities. WCIOQT's along with the Universities' many contributions through medical sciences and the diversities of physics and technologies made nearly every disease nonexistent.

Waste and trash and pollutants were a thing of the past as everything, all materials, were recycled. Enormous trash dumps that once existed outside the Domesticities, were emptied by recycling material into various products by personal and professional three dimensional printers and regenerators.

Humans continued to enjoy the occasional outings to such places as the public cinema although cinemas were common in every individual and community Bio Dome. We however enjoyed the possible anomaly when it came to enjoyment. Such as a pizza that didn't have a perfect arrangement, as our regenerators would produce it. No, on occasions we all seemed to prefer some degree of imperfection. The perfections of proxies although alluring, eventually became unbecoming. What we appeared to search for, we now had, and we didn't pretend to be thrilled with the outcome.

With our goals met, we looked for new goals, and we perhaps counted on the imperfections of nature to provide us with the next hurdle, the next challenge. Death seemed to meet that criteria and Wyatt had many in support of his possible achievement. He ventured on in search of the death of death.

As Proxies, we eventually ventured toward the virtual worlds in our search for some sort of stimulation. Although distant, we searched for what would make us more human, more animal and not machine...and that for us seemed to be found in violence.

Programmers were tasked to produce nearly every war that had existed throughout human history as well as dystopian futures in which was featured totalities of annihilations. This surprised us at first about the Proxies, but then made us realize that our own survivals were based off of historical horrors that seemed the foundation of life. Life was always challenged by death. Death allured us to it, like a moth to flame.

Our Domesticities weren't just to keep us from further destroying the planet, but they were also required to keep us safe from the release of tremendous amounts of bacterium and viruses that were now our worse enemies as they searched for their food, their means of survival, and the human body. Some of the Oceans viruses were in fact mutating, and that was most unfortunate as the Ocean contained trillions of viruses that at one time held no threat to the human.

Wyatt knew, that, he had not yet conquered death. But if he could, it would not control us, and we would no longer be subject to the foundation of survival. We would evolve past life and into a new horizon that was not subject to the definition of the other. There would only be existence, forever, existence. This was Wyatt's goal, as he hated death like no other foe. To Wyatt, death was not our ultimate challenge but our failure.

Wyatt and his teams at WCIOQT worked vigorously at perfecting the Proxies, past what the UC had approved, through the help of The Mother. The Mother would only help with the transference if the person requested they be transferred and their subconscious was in agreement. This many times delayed the transfer and on such occasions there had been loss of life. For the most part however, humans were about to experience the longest life spans in the history of our species.

If Proxied it was estimated to be in the thousands of years, and synthetic parts were interchangeable and available at all hospitals. If one remained human, yet obtained processes of biological and some synthetic transplants, the estimate was a bit lower at four to a possible five hundred years. Either way, the human being began to realize its youth and attitudes adjusted to these new time limitations. Having children became nearly unnecessary. The majority of humans chose to not have children at all, which may have resulted from this removal of threat and no longer focusing on survival and the probability of death.

A new movement toward the gathering of information took hold. Humans accumulated at the Universities, as well in the main dome of the Domesticities. The central areas were always active because it consistently remained day time and the weather was controlled (individual and community Bio Domes were under the control of the residents). This allowed the population to advance in intelligence, and soap box polemics became very popular with several such debates being broadcasted around the world.

Universe City's influence was perhaps the defining moment of our history, as it was then that the world's mindset; if one could consider the entirety of the population as a mindset, was now on the verge of possible change. But how would a once violent species accept a worldview that would perhaps veer our race toward peace? Would our biology revert to what kept it competitive in an otherwise violent world?

Proxies' transference was as advanced as biology could perhaps establish itself at this time. Human's neurons relayed information over to their synthetic and mechanistic partners. Biology was now becoming fully artificial or what we thought was the next evolution of the human being, the singularity as it was termed.

CHAPTER 34: AMANDA

When you are courting a nice girl an hour seems like a second. When you sit on a red-hot cinder a second seems like an hour. That's relativity.

Albert Einstein

Amanda Lewis was now a full time employee of WCIOQT and Drake Caron's soon to be fiancé. Benjamin Mitchell had attempted to get her interest, but Drake was possibly a bit more confidant and persistent. Drake, Amanda and Ben all graduated with honors from medical school. Ben went on to practice general surgery which now incorporated nanotechnology. Amanda and Drake worked side by side in medical research at the WCIOQT labs for many years.

Ben was dashing and a rather well known ladies' man. His thick baby fine hair usually hung loosely onto his shoulders unless he was working, and then he would bundle it into a nylon hair cap.

Surprisingly Ben married quickly out of school to another medical doctor, but unfortunately it lasted less than two years, and that amount of duration only because they were so busy that they hadn't the time to get a divorce sooner.   He'd go on to date others, but he always held a flame for Amanda. What he actually wanted in life was a partner. Someone that could keep up with him no matter the conversation. Amanda usually knew how to tame his curiosities by offering numerous avenues of information and challenge. He at times felt a bit intimidated by her genius.

Amanda was beautiful for the standards of those times, but most of all quick in wit and funny. Her list of jokes seemed endless, and she obviously loved life. No matter the situation, she had a way of making everyone feel good about themselves, and they wanted to be in her presence. Wyatt immediately fell in love with her and welcomed her into the family in hope that Drake would make it official. He'd ponder having a daughter, and how Dom would've loved her too.

On the day that Drake made the announcement, Wyatt was more relieved than surprised. Ben was a little broken hearted at first, but knowing that this meant that Amanda was in the family for keeps made it a little easier. He would always love Amanda. The possibility of a permanent situation however would make her part of their team, and that was somewhat sufficient. Wyatt had called them "The Three Musketeers" and now there was "D'Artagnan!". Yes, it would be complete, and The Musketeers could not have imagined a more perfect union.

Once the media got a hold of the announcement, there was celebrations at nearly all the Domesticities. Celebrations were common, and appeared to have little need in order to hold parties as they were opportunities to socialize. The Caron's announcement however was one in which just about everyone on the planet held a personal connection. There were numerous stories of how their breakthrough discoveries benefitted either themselves or someone they knew personally.

Amanda's parents had died during the Drone War. Her Father was a Colonel in the Air force and her Mother held the prestigious position of United States Speaker of the House. They were both enroot to return home from a visit to the White House when an enemy drone fired on Air Force Three. The airplane was designated as such due to several threats, and therefore the US military could then supply protective support. Our drones along with two accompanying stealth fighters were able to bring down the other eleven drones. Unfortunately, everyone on Air Force Three was killed, as well as forty one people on the ground.

The memories of her loss, made her want a small ceremony. Her parents couldn't attend, and there was no one else left in her family, so her side of the room would be empty. Noriko however, almost kicked in the door in place of Dominique as the coordinator. Noriko was sure that Dom would have filled in as much as she could for Amanda's loss. So, Noriko took the controls and pushed full throttle ahead. Noriko could fill any room...she was after all one of the founding mothers of the Domesticities. Wherever Noriko lectured, it was sold out, standing room only if you didn't manage to get a ticket.

Amanda didn't mind Noriko taking over at all, in fact she was relieved and thankful. Besides, she had more than her hands full with projects she and Drake were working on together. She also pioneered a few small but extremely important innovations toward medical advancements, which she received several Nobel mentions. She was not only Drake's motivation, but his competition at times.

Wyatt admired his future daughter in law's tenacity when it came to committing to any project, as he felt that if ever there were to be grandchildren, Amanda would make an incredible mother. There would however, be no such children for many years to come. Some at this time were choosing never to have children due to humans extended life spans, and the current continuing population of the planet. Nature as well appeared to generate an infertile pattern among humans...many could not have children unless science made it possible. Most simply accepted their circumstance.

Despite that many humans had lost their lives in the war (Proxies weren't considered quite human), our current population remained in the billions, from nearly fourteen at our peak to over four billion currently. Therefore more Domesticities were being constructed in order to house all of the human population. With death becoming nearly non-existent and births still occurring, the need for additional resources continued. However, through technological advancements along with The Mother, we remained fulfilled.

Eventually all humans would move into the Domesticities by law, and almost surprisingly without struggle. Most people by now simply wanted peace, and besides, the Domesticities were incredible masterpieces of aesthetic beauty. The personal and community residential bubbles were also of such amazing beauty and utilization, that humans took to living in such Domes as being in a paradise, any paradise of their choice; as one could program individual Bio Domes to their preferences. The indoor jungles were in the beginnings, but many large plants were incorporated and growing rapidly. Much of Earth's botany had been stored in enormous vats around the world in order to save them from ourselves. We were then able to construct vertical farms throughout the cities, and thereby bring back fresh produce to an otherwise basic regeneration machine's reproductions.

Drake and Amanda wanted a quiet and personal wedding with only their very closest friends, however they agreed to allow it to be televised in order to keep some control over the frenzy. Noriko had the Caron's mansion renovated to include a large ballroom, which would accommodate the celebration. She designed it to allow the floor to be rearranged when it might become necessary for Wyatt to hold large meetings. It was quite a spectacular room with its over hundred foot domed ceiling that offered several surrounding screens that could display varied skies. It reached up to nearly the limit of Wyatt's personal Bio Dome.

"Nori, you sure know how to build 'em." Wyatt watched as the AIOs put up the finishing touches.

"Mucho Gracias, ah...thank you very much...from you, this is great compliment. (Noriko realized that speaking Spanish or even her native tongue of Japanese, would only remind him of Dominique, as Dom had spoken both languages fluently) I'm happy you happy." Noriko smiled widely and clapped her hands then placed her fingers over her lips, as if in attempt to control her smile, but her grin was obvious and her eyes glistened with glee. She loved making Wyatt happy, he deserved happiness in her opinion, and Dom would've done the same.

Wyatt glanced at Noriko, "If only Dom could've been here...she would have had so much fun helping you get this together for the kids."

Wyatt smiled and continued to look upward toward the AIOs working on the ceiling.

"Yes, si...she have done it much moa better." Noriko's face not only took on sadness over the loss of her friend, but of possibly reminding Wyatt of that loss, by her difficulty with English.

Wyatt looked over toward her. "Oh hell, Nori...I'm sorry, I didn't mean to...".

"No, no, no...I think about her too...all the time...every day, my life..".

Wyatt nodded… "She would have loved Amanda."

"Si, mucho…ai…my English! Noriko rolled her eyes, then she continued… Amanda love her too, she would, si."

"Indeed." Wyatt nodded in affirmation. "Your English is wonderful, you're getting better and better. Considering that you're not relying on interface interpreters". Wyatt grinned. "Well, I'm glad that we have you here…I can't tell you how thankful I am to have your help." Wyatt walked over to Noriko and hugged her. "By the way, how's twenty seven coming along?" (Wyatt was referring to Domesticity 27 in what was once Spain) Wyatt released his embrace.

"The AIOs have near finish. I say, maybe one more year, then all done. You should visit?" Noriko waited for Wyatt's reply.

Wyatt's eyebrows gathered, "You're right of course, Nori…I should visit. It always seems however that I am neck deep into something or other."

Noriko nodded…she suspected that Wyatt was continuing to mourn the loss of Dom, and perhaps she only kept Wyatt reminded of the pain. She knew all too well how that torment felt, as each time that she would hear of or see an image of Wyatt, whether reminders that were televised or by a glance of a photo of the past, her heart would feel the now familiar heaviness. When she would see Drake, she would instantly see Dom's face in him, which pulled her to somehow take the roll of mother. It was perhaps why she would drop everything the moment that either one of them needed her.

Noriko while in thought, smiled slightly, "You and Dom raised a good boy, and I'm more than happy to be here for him, anytime he need me, he just call…anytime…and now Amanda too…anytime."

"Thank you, Nori…you're our family, you know that…and the same goes for you…if ever you need anything, anything at all, just call and I'll be there."

"I know mi hermano… I know." Noriko nodded smiling.

Suddenly Drake appeared which startled Noriko. "Oh, sorry Aunty…I need to speak to Dad, if you don't mind?'

"Uh, of course…I have lots of work to do. I let you have him." Noriko responded as she reached up and pulled his head down to her height kissing Drake on the forehead, she stepped back and walked away from both men in order to check on the AIOs progress.

Wyatt looked at Drake curiously.

CHAPTER 35: SURPRISES ABOUND

Man is always more than he can know of himself; consequently, his accomplishments, time and again, will come as a surprise to him.

Henry Wadsworth Longfellow

"The Mother had the AIOs produce the new interfaces. These are permanently ours Dad, courtesy of the UC." Drake handed Wyatt what looked like a twentieth century U.S. silver dollar, except that it had no markings.

"Hmm, rather simplistic in appearance. Do you know what its composite is?"

"I had it analyzed at WCIOQT and it's pretty similar to the original ones that we made...they said it was made up of silver, silicon and a form of diamond that leans more toward graphene layers and carbon nanotubes as the diamond flaw, perhaps more path than flaw...so...anyway...I'm still not sure how that could possibly relay holograms mechanically, but I wanted you to see how they work, Dad...check it out...Albert...Doctor Einstein?...". Physicist Albert Einstein suddenly appeared.

Wyatt seemed stunned as he walked around the holographic image. "This is incredible! Where are the lasers? There's no lasers." Wyatt looked around the ballroom for possible projectors.

"Yeah...isn't it wild, Dad! At first, I did the same thing, until I asked Albert how this was possible and he said, "Through same space." He explained that entanglement was something like what we refer to as wormholes, but that when particles are entangled they actually exist in the same space. So what we are seeing as holographs are The Mother's ability to bring any form of projection into the same space. It doesn't need lasers...it just needs lasers to exist somewhere. Isn't that amazing?!"

Wyatt nodded, "I expected some wild and amazing discoveries, and The Mother is certainly not disappointing me."

"The Mother probably knows who you're gonna pick, Dad...but The Unified Coalition requested that each person assigned to an interface must choose their holograph as a validation for The Mother. Jaiobian should be contacting you with the details, as they've written up a bunch of guidelines for the QC (quantum computer)."

Drake looked toward Noriko as she spoke to one of the human supervisors of the AIOs. "Sorry to have interrupted you and Aunty Nori. But I wasn't sure if the UC is awarding her an interface and I didn't want her to feel uncomfortable."

"I understand, Son...good call. I can't imagine Nori not getting approval...all she ever does is think about what she can do for others. This Domesticity wouldn't exist if not for her and your mother."

Drake nodded, "I agree. You may want to ask Jaiobian."

Wyatt looked down at his interface. "Hmm".

Drake chuckled". You're not thinking of what I'm thinking, are you?"

"Doctor Richard Feynman." Wyatt spoke toward the interface. Doctor Feynman appeared in hologram in front of Wyatt.

"How can I be of service?" Doctor Feynman asked in his familiar East Coast accent.

Wyatt was a little taken back to when his father spoke fondly of the days that he spent with Doctor Feynman shortly before he died. "Uh, yes...will Doctor Noriko Tomita be awarded an interface?"

"Probability is that she is on the first list along with you." Answered Doctor Feynman.

"Thank you, that will be all for now." Wyatt watched as the hologram disappeared.

"Dad...oh dad... you are so bad, no points for you!" Drake laughed.

"Hey...I was simply validating, hmm... maybe worth a point or two?" Wyatt winked.

CHAPTER 36: LAW

Rightful liberty is unobstructed action according to our will within limits drawn around us by the equal rights of others. I do not add 'within the limits of the law' because law is often but the tyrant's will, and always so when it violates the rights of the individual.

Thomas Jefferson

Wyatt's cell phone rang, and Jaiobian was on the screen. "Hello Speaker Richards. Drake said that you would be calling."

"Do you have your interface with you, Wyatt?"

"Yes Speaker...in fact, I just recently validated it by choosing Doctor Richard Feynman." Wyatt grinned.

"Good...did it work sufficiently?"

"Yes, and immediately. Quite an incredible likeness."

"Very good. So far, all interfaces are working. The only thing that we have to monitor them is the system itself. The system, I mean... The Mother, appears to be up and running perfectly. We're currently in deliberation yet again over the guidelines, but for now Wyatt, I suggest that you and Drake play around with it for a while and let us know if you find any glitches. We also have interfaces for Doctor Noriko Tomita, Doctor Benjamin Mitchell, Doctor Shen Lum, Doctor Trent Keamoku, Doctor Leonard Star, and quite a few at WCIOQT. Oh, and Amanda should have been given one by now. It was given to Drake to pass along."

Drake nodded, "Yup, I'm here too Speaker...I gave her it right before coming here to give Dad his."

"Oh, there you are, Drake. Good, good, good. It's been a bit chaotic trying to get things organized, but The Mother has been more than helpful, and we should have just about everything in place in a few hours. It would probably have taken years, but with The Mother, anything we ask of it seems immediate. It's already doing stuff that I guess I would have requested of it eventually." Jaiobian smiled.

Wyatt nodded, "It certainly is extraordinary how this quantum computer appears to be interconnecting with our specie. I suspect that it's been around for quite some time, or at least its potential."

"Well, it's been a popular theory that some sort of fundamental field toward consciousness may exist." Jaiobian responded.

"Yes. The potential field, by reason and logic, should. There also may be a means toward opening perhaps Universal communication if such a language exist. However, we could also be opening a paradoxical box in that some species or other intelligences may consider our attempts at communication as possibly detrimental to their existence." Wyatt retorted.

Standing behind Jaiobian, Jose offered his question to Wyatt. "Hmm...Do you feel that The Mother may bring us enemies or possible detriment, Doctor Caron?"

"Not necessarily Representative...however, we should perhaps be cautious." Wyatt offered.

"Hmm...Yes...perhaps we should." Jose nodded while scratching his chin.

Jaiobian appeared concerned, "Well then, Doctor Caron, we may want to put some sense of restraints on possible communication at this time?"

Wyatt nodded in affirmation. "Perhaps, yes."

Jaiobian again spoke, "We certainly must consider our safety. The war has taken so much of our morale, that any threat to our species at this time, may give rise to total rejection of The Mother. We stand to benefit a great deal, however, we must continue to regulate this new power, in my opinion. Such must go before the entire Unified Coalition. Many amendments to any established constitution under this system must be written. One can imagine that The Mother has already written the documents in the UC hall of records." Jaiobian grinned.

Wyatt again nodded, "I'm sure that it has it waiting for the UC. Perhaps we should talk again after the UC is aware of the structure of the new system?"

"Yes, we'll get back to you, Doctor Caron. We had thought that we had a handle on this, but obviously we have a lot to learn. Bye for now." Jaiobian closed her (thanks to The Mother) now holographic screen, as Wyatt acknowledged and closed his end with a wave of his hand.

CHAPTER 37: WEDDING BELLS DO RING AGAIN

After all these years, I see that I was mistaken about Eve in the beginning; it is better to live outside the Garden with her than inside it without her.

Mark Twain

The ballroom had been completed, and the limited guests' list had been assigned to Drake and Amanda. Aunt Noriko had refused to do the list as to appear biased, and besides, it would be up to the discretion of the bride and groom whomever they preferred to bear witness.

The discrete list eventually took on an alarming number. Amanda couldn't bear to leave out any of their friends and associates, of which was basically all of WCIOQT, or nearly sixty thousand in Domesticity 11 alone. Drake continued to remind her that it would be televised. Amanda however was concerned as to whether anyone might feel snubbed and offended.

Eventually, Wyatt and Leo, along with Ben and Drake as a force of one would convince her to agree to a very small ceremony followed by large parties held at not only Domesticity 11 but many of the other Domesticities. Compliments of the Carons, although Amanda having earned a considerable amount of points on her own tried to insist on paying for part.

Weddings were becoming almost a thing of the past, as the need and want to have children diminished almost abruptly. With humans now living longer, even undeterminably longer such that could possibly be in the hundreds, and perhaps thousands of years if Proxied, many former requirements were no longer necessary. We no longer had need to slaughter animals for food as the regenerator machines as well as three dimensional printers were of such high caliber utilizing atomic information, that one simply downloaded recipes and identical nourishment to the image that was provided and thereby was formulated for consumption.

Humans however and on occasion did seem to prefer going to local restaurants even though there was little need to travel far as the centralized University offered most entertainment avenues. Servers at such establishments had only recently been completely replaced with AIOs, however, those humans that chose to remain in supervisory positions, were now under the new point system to which The Mother was now rewarding points toward contributions to society.

Noriko realizing that most humans enjoyed the unpredictable had made the arrangements to have a few of the local restaurants cater the event.

Ben was ambivalent, not quite pleased of the situation as everyone appeared to be, however he was happy for his best friends. His apparent melancholy in the realization that he would never be suitor to Amanda, gave him a degree of guilt that he fought to shed. He would however be Drake's best man, and besides, Amanda did indeed have several friends that Ben found to be interesting.

Amanda almost felt sorry for her bridesmaids as Ben was very charismatic with a keen ability of persuasion. His good looks and genius as well as genuine passion toward altruistic goals, made Ben extremely popular amongst many possible partners in Domesticity 11. It may have also helped that he was considered a Caron. It was therefore Ben's choice to remain untethered.

Besides, his marriage during Drake's and Amanda's courtship didn't last long due to his dedication mostly to his career. His passion was obviously for medicine as much as Wyatt's, Drake's and Amanda's was for medical nanobotics and subatomic theoretical physics. However, the Carons did venture into many branches of science. There seemed no limit to any of their curiosities.

CHAPTER 38: THE FUTURE LOOKS PROMISING UNLESS

If you think in terms of a year, plant a seed; if in terms of ten years, plant trees; if in terms of 100 years, teach the people.

Confucius

"Wyatt, we may have a problem." Speaker Richards announced with concern.

"Is this something that the UC or The Mother can't handle, Speaker"?

"Yes Doctor Caron, or I wouldn't be bothering you. I'm extremely sorry for this interruption, we are all well aware of the up and coming wedding of Drake and Doctor Lewis."

"Go on Speaker...how can I help"?

"Well, we're hoping that you can use your clout to fix a recent situation."

"Go on'?

"We had decided to relocate the prison holding facilities to Earth, and the prisoners on Mars were being brought back and were supposed to arrive today. Unfortunately, Wyatt, there has been a crash, and the bodies are being flown in the new suspension units to Domesticity Eleven's air arrivals and then to Hospital One, where as you know is where Doctor Mitchell is assigned."

"That's easy enough Speaker Richards, simply request Ben. But I'm curious as to how the Mar's spaceship crashed? I'm sure that The Mother gave the correct coordinates, in particular as they reached the space elevator's glide path.

"We haven't received all the details, but suffice to say that we have some suspicions. The Mother has relayed the events that had taken place, but they are somewhat ambiguous. There was an apparent disruption or fight once the prisoners were awakened. I'll keep you informed."

"Thank you, Speaker, I'd appreciate whatever updates come your way. I can't imagine why they would create havoc, their stay on Mars was uneventful. So much so that their engineering expertise was put to good use. I had thought that it was one of the main reasons that was used in deciding to relocate all holding facilities in each of the Domesticities?"

"Yes, Wyatt, you're correct. It was an easy decision based on the fact that we now have The Mother, and can keep track on all activities."

"How was the crash ambiguous?"

"Ontologically each individual has their own rational way of thinking, Wyatt. And that leads me into my next explanation about Doctor Benjamin Mitchell. The Mother didn't allow the assignment. Ben's subconscious appears to be removed from helping the prisoners."

"That's certainly not Ben's style at all, Speaker. He's one of the most compassionate young men we've ever known. He's like a son to me."

"Yes, I know. But this is what we're getting for a probability from The Mother. So, instead of jeopardizing anything, The Unified Coalition has decided to request that you talk to Ben about this very sensitive situation. I fully realize that it may be because these are the particular people that sabotaged Universe City, but we currently don't know how far their connections run and we are trying to detour a possible terrorists retribution. The Mother has given us the probabilities. It will not however give or maybe divulge a non-threat. We've speculated that might be because there is always some amount of threat between groups of varied ideals. But so far it's not yet a threat and we'd like to keep it that way."

"All right Speaker, I'll have a talk with Ben, and I'll get back to you."

"Thank you, Wyatt. We all appreciate this tremendously. Doctor Mitchell is the best when it comes to this new nano-medicine technology, and The UC immediately voted to get him on it. Talk later, bye for now."

"Good bye, Speaker." Wyatt called Ben and suggested that they meet at the Caron mansion which was now located in an underground carbon nanotube bubble built by AIOs. Ben arrived within a few minutes via his gift from the Carons, an antique Maserati Granturismo that was now equipped with a new bio-friendly engine.

It was a far cry from the newly designed public transportation bubbles that would be operated by AIOs built into the vehicles and assigned to each individual. Such bubbles would run strictly on lanes designated for unlimited speeds in excess of 200 mph and driven by AIOs. Ben would have to and did drive his vehicle on lanes for manual operation by humans. Many humans demanded such lanes be made available, although no accidents were to ever occur under AIO operation.

Ben met Juan (Wyatt's first AIO) as he drove under the Porte Cache of the Caron mansion. Juan opened his door and welcomed him into the great room where he could see Wyatt approaching. "What's this about, Wyatt?" Just then Ben received a message from Hospital One.

"That's probably a message from the Hospital, Ben?"

"Uh, yeah, you're right...how did you know? You getting psychic, Sir? Or, did your holograph tell you?' Ben laughed as he looked around the room to see if Wyatt's holograph of Doctor Richard Feynman was on. "Uh, I gotta get this, Sir...just be a second or two I'm sure." Ben began to speak to the administration. "Yeah, okay, sure...I'll be there in a few minutes, if they arrive before I do, keep them in suspension, I'll personally bring them out. Let Taylor know to start the nano-scanner and hologram projector and get them ready for correlations. Okay...yeah, I'll be there soon." Ben hung up his cell phone. "So, what's up?' Ben looked at Wyatt.

"I was called by the UC on this, Ben."

"Why?"

"These are the prisoners from the Mars Prison, Ben."

"Yeah? And?"

"They were the ones that sabotaged Universe City."

"So they're assholes, Wyatt...but I'm a medical doctor that in probability can save these assholes...so what?"

"Well, as you're aware, The Mother is needed for the nano surgeries, and apparently your subconscious is refusing to partake in the surgery."

"What the hell?!"

"Uh, yup...we have to figure out how to get both of you on board or The Mother won't budge, and the hospital may have to call someone else in to perform the surgeries. Thing is, you're the best that there is, Ben, and this is so sensitive that The UC wants to be sure that the best is on it."

"I guess I can understand my subconscious take on this, because why the best for assholes? However, I could guide another doctor or team?" Ben asked as Wyatt appeared to be in thought.

Wyatt blinked, realizing that Ben was waiting for his response, "The UC can't look like they intentionally let them die by getting someone that wasn't top notch. Hmm, well, I'd guess that The Mother wouldn't allow you as your subconscious would probably be against that action as well."

"Can't we override Mother, Wyatt?"

"This is one of those touchy areas, Ben, which was voted in favor of by The UC. We didn't want The Mother to be too much into our lives, remember that we wanted choice, free will, to include not overriding our subconscious decisions that can happen quite a bit before our conscious is even partially aware. We've known for many years that our awareness is rather slow."

"What do you suggest that I do, Wyatt? How can I get control of myself?"

"Well, I'll take us to the Hospital and we'll discuss it on the way. As it currently looks, it may be difficult to find any physicians that don't hold onto some form of discriminative bias toward these people. And quite frankly Ben, you are in fact the best we have."

As Wyatt and Ben approached Wyatt's SUV that was fully equipped with the latest bio-friendly engine, Wyatt's AIO driver also approached the vehicle. "You won't be needed for this trip, two one seven, I'll be driving, thank you."

"Very good, Sir." The AIO responded nodding and then turned and walked back toward the garage door, where it had been previously standing at attention.

As Wyatt readied himself for his drive to the Hospital, he activated his hologram by speaking to the small disc interface, "Doctor Feynman?" Physicist Richard Feynman appeared in the back seat, and Ben jolted slightly.

"Sorry Wyatt...even when mine appears, it sometimes gives me the willies...they look so damn real."

Wyatt nodded, "They do at that. I like to sometimes pass my hand through them as when The Mother incorporated nanobots into the projection it gave the holographs a sense of something being there." Wyatt grinned.

"Okay, I haven't done that, yet, but in my opinion, that makes them even creepier." Ben rolled his eyes.

Wyatt chuckled as he got into his vehicle and looked in his rearview mirror at his holograph. "Well, The Mother sure did make its holographs amazingly realistic, our originals look like a child's attempt." (Ben nodded at Wyatt's remark, the whole time in a stare analyzing the holograph of Richard Feynman.) As Wyatt continued to look at his rearview mirror he wondered if The Mother would pick up the request from a reflection, "Uh, okay, here we go...Richard, please inform us if there is anything that we can do for these people that have been injured? Will you allow such?"

The Mother as Doctor Richard Feynman answered immediately, "Yah shuah...two physicians are being brought together that will be supervised by Doctor Amanda Lewis. I have contacted all members with your instructions."

"Thank you, Richard. That will be all."

"Damn!" Ben shouted.

"Damn amazing!" Wyatt bellowed.

Both men laughed, then sat back in silent contemplation as Wyatt turned onto the pavement of the Hospital entrance.

Wyatt shook his head as they slowed to a stop..."it's a good thing that there wasn't any traffic...I really can't remember how I got us here."

CHAPTER 39: ODDITY OF DEATH

Death is a very narrow theme, but it reaches a wide audience.

Socrates

Doctor Amanda Lewis met Wyatt and Ben at the entrance. "Geez guys, I'm glad that you're here...this sure the heck isn't my specialty and I'm about as rusty as the nails on the Flying Dutchman."

Ben chuckled and attempted to make her laugh by speaking like a pirate, "Argh,well preserved nails there matey! Argh, well preserved mind you...now quit that there worryin', ain't time fur that, it'll come on back to you, argh, once you get in the captain's chair."

Amanda grinned closing one eye, putting her hands on her hips, then swinging her left arm and fist back and forth in front of her torso in an attempt to respond as a pirate, "Argh, am I expected to guide the other two physicians?"

Ben smiled and then took on a serious persona, "I think that it would be better if all three of you coordinated the manipulations. If any of you have any sort of reservations, The Mother may shut down the operations, and that can't happen. If we can keep The Mother involved, they all should survive, given their own acceptance to The Mother's interventions."

"Well, okay, I guess that in a way I really don't have much choice in this matter...pardon the pun." Amanda said with obvious sarcasm.

"Yup. From what I understand, The Mother was only able to find the three of you from this hospital. Apparently your subconscious minds are willing to save these people. Mine unfortunately has some issues." Ben shrugged. "Would be nice if the guy would talk to me about it."

Amanda shook her head, "Yeah, ever since The Mother has become part of our lives, so many have been saved from future detriment, but it's so weird and strange. It feels as though the world is watching us. Like some astronauts would report the sensation of being watched out beyond our planet, but now it's in our homes, our backyards, walking down the paths in the main dome, it's all around surveillance. So, uh...sorry to ramble...let's get this going. Do I scrub up these days?"

Ben shook his head, "Nope, not anymore. The Mother will mist you and clean off all bacteria, viruses and any fungus among us." Ben smiled at his retro pun.

"Hey...I may disappear!" Amanda responded.

Ben laughed as they walked through the disinfection tunnel toward the surgery rooms. Laying in two rows, were eight bodies floating in suspension."

"Oh my, there's quite a bit of damage to each...but I don't see anything immediate yet. Mother please open individual body holographs", requested Amanda. Instantly eight holographs appeared of the internal organs of each patient. Amanda motioned her arms and the holographs expanded as she walked through the images, commenting on the damages and procedures that would be necessary. "They were lucky, basic organ replacements, no paralysis, no immediate threats, hmm, one and five will be the first two, hmm, and six...yes, those three...then the rest."

Ben whispered to Wyatt, "That's our professional." Wyatt nodded. Drake walked up to the two men. "How is she doing?" Ben jolted, "Sheesh Drake, try announcing yourself next time."

Drake whispered, "Oh, sorry about that."

Wyatt whispering answered, "As soon as we went through the tunnel, she took the controls."

"That's our D'Artagnan...and the love of my life, Dad." Whispered Drake.

Wyatt smiled and patted Drake on the back.

The two other physicians arrived and immediately Amanda passed along all the information that she had so far diagnosed. Orders were given for organ replacements, however given the time it took to get these humans to the hospital (nearly six hours due to the ocean crash site), The Mother was fully aware of which organs would be necessary and they were awaiting transplantation.

The Mother began the repair procedures as directed by each of the surgeons via the robotic surgeons that the three were now hooked into. The robotic surgeons would perform the actual surgeries while the human surgeons monitored.

Ben, Wyatt and Drake monitored the procedures from remote viewing positions above the room in which the walls and ceiling were screens, thus the procedures were shown in a 360 degrees arena. As each of the organs arrived, they were scanned and then placed into the cavity very quickly, so quickly were the operations done that any amount of further trauma was nearly non-existent.

An hour later, five of the surgeries had been completed. Three remained in surgery for another two hours as they had suffered severe burns which required skin grafting, which the robotic surgeons produced via printed synthetic layering. Although application was rapid, there was a following application of protective coating that would guarantee acceptance with little chance of rejection or infection.

As the men watched the procedures being completed, Drake spoke, "Dad, do you think that we're doing the right thing by saving these asshole's lives?"

Wyatt looked over at Drake curiously as he spoke, Drake's head was angled downward and his eyes were fixated on the medical procedures displayed on the surround screen. "What do you mean, son, these are human beings?"

Drake continued to follow the procedures, "They were perfectly willing to kill millions of children at Universe City. Why? Because their particular oligarchy didn't like the idea of...what? That these new breed of human beings might save the planet? Might make us get along, unite, care about one another instead of focusing on ourselves? See, that's what children show us, that it's not all about me myself and I, it's about them, our future. Getting us to that next point in evolution. But no, of course not...we can't do that! We must continue to horde and focus on ourselves because that's the way that nature intended us to be! Cruel, self-centered, greedy, narcissistic rule the world in order to sit on the thrown mentality! Yeah, that mentality really worked out for us didn't it? We nearly destroyed all of life on this planet (Drake's voice dropped to a whisper) including ourselves, you fucking idiots."

"I realize that these aren't the sort of people that we would choose to emulate, Drake. But they were somehow convinced that what they did was righteous. So in their minds, they were doing it for what they believed to be good. Your Mother and I raised you to realize the relative nature of good and evil, benefit and detriment. It's sometimes just a matter of which side of the fence one chooses to stand on."

"If we don't fight the bullies, Dad, than who does? Do we just let 'em hurt and kill children?"

"No, not at all, son. In fact it gets easier from here on, as now we have a quantum computer, that will know the probabilities before they happen, and we can perhaps stop most crimes."

"Oh, Dad, you of all people know that's a slippery slope. We can't prosecute anyone unless a crime has been committed. So far, all the police do is give out warnings. The Mother can by theory give us the actual future...not the probability but the actual future. It should by theory be able to time loop, but no...the UC doesn't want that."

"Yes, that would theoretically be true, but that's the deal, Drake...The Mother will know it before it might happen. Actually by theory, The Mother knows the past present and future in each moment of entanglement. Its superposition state, should by theory open it to every reality. It's not just probabilities that we're dealing with, it's a system that is entangled to time. But do we really want that? Do we really want to know the future?"

"So, basically Dad...it's a god? I thought that we finally got past that nonsense? And now, we discovered it, and we don't want to know what it knows?"

Wyatt laughed, "No, I wouldn't say that it's an absolute. We may have triggered the entanglement...maybe others before us did the same? Maybe not, I don't know. Maybe it's a foundation principle of this particular universe? There's many questions, which we may now be able to answer. But if it's only one universe or one in a multi of universes, it may only carry the information bordered by laws of one particular universe. Currently it's this one."

"If you think about it Dad, if it is entangled in the past present and future, than what's the purpose of our existences? If it knows everything, than we're irrelevant?"

"That would be true I suppose if time didn't exist, Drake."

"Huh? But time is irrelevant to entanglement, Dad. Time is at a point of past present and future, as one, all happening in some sort of now. So, we're irrelevant?"

"We do however exist, Drake, in those diversions of time...so we are indeed relevant. We are ruled by time in a sense. What this does suggest to us is that these information points exist in order to formulate time lines else entanglements wouldn't exist...or it might be reversed, no one can be certain. This could be likened to a tapestry of what we call existences weaving in and out of time. What we have come to call reality. We are real, we are in fact reality. Infinite points of existences, in which each point represents a totality of information up to that point. Put each of those points together and you have the bigger picture per se'. Does The Mother hold the bigger picture? I'd say that given infinities, it holds it as it's being painted. It completes each painting instantly by filling in the most likely probabilities, or as you're suggesting by knowing. But we are possibly the artists by our own choices."

"But if our choices are known, Dad, how could we have choice? Free will would not exist, it would be an illusion?"

"Well, son...it would in fact exist, The Mother doesn't necessarily know, but it can and does generate the most likely probabilities. Just because it knows doesn't mean that it made it so, we did. And if the other theory holds that it basis its best guess on highest probabilities, that makes it appear to be in possession of all knowledge, but it's actually the most likely knowledge. Therefore, probably, and... not certainly. As those points of totalities are entangled to our choices at each moment in time...we are the ones painting, creating the picture per se' as it is being recorded. Either way, it depends on us, depends on free will. The Mother can get as close as the next best guess of what will be painted even if it knows because there's a little fluke that we call anomaly and that keeps any at bay, such as totalities, and...infinities are even subject. The human being is similar to The Mother in that we also can work on probabilities, but we can work also with anomalies. We can actually take chances against the most likely probabilities. Such that we can bet that something might happen even though it most likely won't. I don't believe that The Mother has that capability. It's the most accurate system that we have available, but it may get the answer wrong given that this Universe is based on uncertainty. The human being however, is prepared for possibilities such as anomalies, and, or wrong answers. We adjust, and we make decisions while we adjust. We've known that such events as photonic duality is simply uncertainty, and we simply adjust.

If the Multi-verse theory proves to be correct, then every probability that can exist does exist. What we therefore believe to be uncertain may in fact appear determined. However, quantum physics suggest uncertainty, therefore, the Multi-verse theory may incorporate such uncertainties, and that leaves us with every point or probability existing and also not existing, which opens up the door again to anomalies, which reverts back to humans' ability to cope with those anomalies.

But the really interesting existential question is not only why every point exist and continues to exist, but now that we appear to be on this track...where is it taking us? Where exactly will we be willing to go via The Mother?"

Drake mumbled, "I agree, Dad...and believe me, I've thought a lot on the same lines that you have. Like, now that our law enforcement has the backup of The Mother, including our AIOs, that can respond twenty four seven and are bullet proof and tamper safe by their own regeneration and defenses...the ball is in our hands. The hands of those that choose to be responsible law abiding citizens. Laws that will never be absolute cut in stone, laws that we as humans make up, but are always subject to challenge and possible change, but laws that help us to live with one another in peace. And people like these assholes will be able to see themselves through the eyes of each of us in a sense. The Mother could do that sort of thing by theory...if only the UC would allow it?"

Wyatt didn't respond but continued to listen to Drake and Drake continued to talk a bit louder perhaps because of his passion...

"Once the human mindset begins to change, once each of us is aware that secrets are a thing of the past, we should begin to experience what we as a species have uniformly claimed to want, and that's... peace, true peace. A world in which humans will no longer suffer the cruelties of nature such as that of diseases, mutations or inflictions by predators including one another. A world in which there is no hunger, and little, to no, need. And perhaps because of Universe City, a world which considers children our greatest resource."

Wyatt smiled.

Drake looked up at his father, "Well, I know, it all sounds too good to be true."

"There is sure to be bugs in the system, but with all that humans have gone through, I think that we'll be able to handle the bugs." Wyatt patted Drake on his shoulder.

"I hope so, Dad, I really do hope so."

Ben spoke, "Don't mean to disrupt the polemic but it looks like the surgeries are nearly completed. It should take a few more hours for the skin grafts, but for the moment their vitals look stable."

CHAPTER 40: HUMAN COMPLEXITY

The theory of evolution by cumulative natural selection is the only theory we know of that is in principle capable of explaining the existence of organized complexity.

Richards Dawkins

Speaker Jaiobian Richards was heard at the other end of Wyatt's cell phone receiver. Wyatt excused himself from the observation room and went into the hallway.

"Yes, Speaker Richards, so far so good."

Jaiobian nodded and then responded, "Excellent. The Mother said that probabilities were favorable toward their recovery if Amanda was brought in."

Wyatt nodded, "Can't say that surprises me." Wyatt grinned. "You said that there is some amount of ambivalence in regard to what happened?"

"I suppose there will always be some amount of ambiguities when we deal in probabilities and not absolutes." Said Jaiobian.

"Agreed." Wyatt responded.

Jaiobian continued..."However, The Mother did supply us with what looks to be the instigator of the crash. A young man named Nemo Utis. But this is where it gets tricky because why would anyone choose to die these days? They also have long two hundred years sentences, but that by probabilities is just half of their lives expectancy these days. It once was a life sentence, hmm, we might need to change that. So, anyway, with some reservations and quite frankly a lot of suspicion, we have sent extra security to the recovery rooms. It's easier to escape from a hospital than it is from our new AIO guarded judicial cells. Although, it's tempting to see how The Mother benefits our security."

"Ah, yes." (Wyatt grinned) "Well, although tempting, we perhaps shouldn't evoke The Mother unless we actually need it. I should have guessed that this might have been a ploy in order to get them into a more convenient place to escape. The extra security was a good idea, and I'll pass the information along to everyone here that's involved."

Jaiobian continued, "We're going public with what information we have, and we're about to launch a few drones around the hospital."

Wyatt responded, "That shouldn't be a problem. The public is rather familiar with the drones that are used for the vertical farms and maintenance of the domes. As well as the few that the hospital itself uses for transport these days. But yeah, it's probably best to let as many know what's going on."

"Yes, we thought such, and besides, with The Mother, there's very little chance that anyone can hide. It should therefore be senseless to any of them to try to run. In particular with the AIOs standing guard."

"Indeed." Wyatt nodded in agreement. "The AIOs won't harm them unless they try to hurt others, but the AIOs only act on the defensive. However, do you feel that this might be a bit too soon to completely rely on The Mother? Security with the drones was going rather well with the former gamers, but with The Mother it may remove the stress completely and they may repress?"

"You mean that they may get bored?"

Wyatt nodded, "Yes, exactly. You probably remember when we moved the gamers into security positions?"

"Yes, I do. We didn't know exactly where to place them once the war ended. Security seemed the best option at that time." Jaiobian responded.

"Well, they have their moments when they have to use some of their training, but not many. So, once The Mother is incorporated...then what?"

Oh, yes, I see your point, Wyatt. The police may be necessary for physical arrests, but probably not the gamers." responded Jaiobian.

Wyatt continued, "We're still working out the bugs in the virtual systems. It's noted that the programmers are being pushed to establish highly violent and sexual programs in order to circumvent actual desires. They're even installing full body interaction. With the aid of The Mother, the virtual realities should, by reason, be very real... to the human brain. It may in fact prove to be of benefit to prisoners. We could give them whatever world they choose to exist in, even ones in which they are the Dictator inflicting whatever horror they choose."

"That would be rather counterproductive, don't you think, Wyatt? Not only to the prisoners, but to anyone such as the gamers?"

"To this world in which we live, yes. I feel that we need all these amazing minds in order to contribute to this reality. You know that, Jai, I mean Speaker." Wyatt rolled his eyes at the mistake.

"Oh Wyatt, please call me whatever makes you comfortable...the days of the classes have long since passed, and I am always humbled in your presence." Speaker Jaiobian Richards bowed slightly.

"Thank you, Speaker. I will continue to give you the honor that you deserve. You earned the title by the love of the people. I cannot imagine a greater bestowing." Wyatt bowed slightly.

Jaiobian bowed in return.

Wyatt continued, "But as of yet, we simply can't quite circumvent previous programming of the human mind. And perhaps that is for reasons of that particular person, their own personal ontology, their free will, their ability to be anomalous, unpredictable. In other words, by their choice they may want to behave detrimentally to this reality and perhaps we should be prepared for that, and offer them the option of doing precisely what they would prefer to do...but only virtually? Our gamers would therefore become necessary to add their abilities to be unpredictable at times. They could enhance not only the prisoner's virtual worlds, but that of the Proxies? The Mother in this regard might learn through us as we learn through it."

"You bring about an interesting suggestion, Wyatt. Their own Viking or Spartan heaven per se', that's away from the actual world in order that we can have some form of order, hmm. The Unified Coalition and I will be meeting today, and I'll of course bring that up. We do know that our Proxies so far do tend to play a lot of video games. The programmers just can't keep up with their demand. Our current virtual realities are also no match for their intellects. With The Mother on board, and a mixture of our gamer's possible anomalous contributions it should become very interesting to see how far they can advance."

"My bet is on The Mother, Speaker. I think that a quantum computer can indeed program significantly difficult challenges, and the gamers will enhance that ability. The gamers can in fact earn points keeping up with both The Mother and the participants. It should fill a need, and the gamers may feel needed once again."

Jaiobian nodded, "My bet is that you're correct, and I'd also bet that The Mother should be consulted on that dilemma."

"Good idea. I'll open my holograph of Doctor Feynman here shortly. We'll speak again in a few hours...Sp...Speaker." Wyatt grinned.

Jaiobian winked at Wyatt, "In a few hours, goodbye for now."

"Bye for now", Wyatt smiled and then closed the shield of his cell phone. He looked at it and grinned..."Old Girl, you're now a relic...I should pick up a new cellographone...but I'm rather sure that's what Drake got me for the holidays." He sighed at the idea that the world was changing faster than perhaps he could.

CHAPTER 41: RUNNING

It is not possible either to trick or escape the mind of Zeus.

Hesiod

"Who is Nemo Utis?" Wyatt asked his hologram of Doctor Richard Feynman.

In his rich East Coast accent, the holograph of Doctor Feynman spoke, "He was born to DeSamuel and Carmela Utis on April 29th 2015. The three hundredth and thirty first child to be born in what is now called Domesticity number one before its completion on what was once considered the East Coast of the previous United States of America. The place of his birth is now located underwater, the water that was once used to protect the construction crews from nuclear fallout. The City's completion would eventually require the removal of water, however records show that his place of Birth is in Domesticity number one."

"Okay, okay...I realize that you're called "The Great Explainer", so this could take far too long, uh, I guess what I really want to know is, what caused him to go astray?"

"Astray?" The holograph of Doctor Feynman appeared confused.

"Yes, from our current values within our society. Why did he choose to bring harm to Universe City's residences, the people, and the children? Was he paid off, was it for money?"

"Yes. He was motivated by the monetary reward. He agreed to be paid two million dollars. He was to be eventually paid an additional amount during that time. Five million dollars to be specific. Due to Universe City's means of economic points system, the monetary amount was increased in order to terminate the possibility of a decrease through points." Doctor Feynman waved his hand in what was commonly seen to be his means of expression when he was alive.

"I see, so he felt that he would have a significantly less amount once the transference to points took place?"

"Yes."

"Was this what motivated the others?"

"Yes."

"Who was behind the plot?"

"Probability suggest that you mean the plot to destroy Universe City?" Asked the hologram of Doctor Feynman.

"Yes, that's the plot that I meant. Sorry about that." Wyatt mumbled to himself..." I keep forgetting that you're not just a hologram." Wyatt shook his head at the realization of a quantum computer. "You look so incredibly real. I have to remind myself to think in probabilities when talking to you."

The hologram of Doctor Feynman answered immediately, "The organization called "Free Will"."

"Do you have their ...?"

"Yes."

"...human identities?" Wyatt realizing that the Mother picked up his question from his subconscious neuronal relays, consciously finished his question vocally.

"Has the Unified Coalition requested...?"

"Yes."

"...this information?" Wyatt smirked, "We're far too slow for you, Mother, and we will have to tweak ourselves somehow, in order to keep up. Are there future probabilities of attack by the organization called "Free W..."?"

"Yes."

"Can you please ca...?"

The Mother put the call through to the Unified Coalition.

Wyatt thought about what knowing such information might potentially mean. Should they or even could they arrest and prosecute these people based on probabilities? He knew that the answer was in fact "no" according to the current legislative and judicial requirements. Would "warnings" be enough? "We know what your plans are, so don't do it!" He pondered and then felt a little silly at his own paranoia..."come on now ol' man...we now possess and are at the very least being helped by The Mother, so by reason, we perhaps have nothing to worry about. The Mother will monitor Free Will's member's movements and actions, in keeping the UC one step ahead. We'll know what their next move will be. The Mother knows the future, or does it really? What if it's wrong? We've noted a two percent probability error. How can any anomaly be actually factored in as if it were absolute...it can't. Crap!"

CHAPTER 42: PUT TO THE TEST

In defense of our persons and properties under actual violation, we took up arms. When that violence shall be removed, when hostilities shall cease on the part of the aggressors, hostilities shall cease on our part also.

Thomas Jefferson

"Apparently members of the "Free Will" terrorist group, sadly were made up of several of our Gamers, in fact a few were from the elite group of the Final Two Fifty.", reported Irene Tagalicod as she shook her head, via the B.N.R. (Breaking News Report) station, displayed on the wall screens throughout the cities. The three dimensional features made it appear to be at the location. Translated in every language in use around the world.

Irene Continued, "I know that it's difficult to believe that some of our finest would turn on us, but we were ready for them, folks. They tried to take down the security drones and AIOs that were established around Hospital One here at Domesticity number eleven. But extra security was standing by ready for their plan to take effect. Police along with security AIOs moved in on them and several arrests have now been made. They will go before members of the UC in a few hours. The UC apologizes to anyone that was slightly delayed in that area of the city. Traffic is now flowing properly and no delays should be experienced. We will be broadcasting live, of course, so stay tuned." Irene smiled as the screens returned to the outdoor holographic sceneries being broadcasted live from all over the planet.

The main dome could also project night skies of stars more clearly than the human eye could see if actually outdoors. The central University area was kept in daylight...while much of the surrounding areas were programmed to show the night sky in order to give some regulation to our biological systems. When it was daylight outside the Domes, the entirety of the Dome would show daylight. Dom and Nori made sure that the designs of the Domesticities had nearly every imaginable advantage, including systems in place to clear out the weather in the domes and distribute moisture and oxygen elsewhere.

They however, didn't plan for any possible outbreak of violence except perhaps the enormous surveillance system, mainly in place to keep track of mechanical operations, not humans. Or, at least that's what they had hoped that it would only be needed for, but they unfortunately were wrong.

The potential toward violence appeared to be a trait ingrained into biological systems. The fight for existence. It was in our games, in our written works, in our movies, in our virtual realities. When we dressed for the stage, the world, we played whatever we felt that the character required. Our interpretations of the parts of the play we portrayed as best we could, we were a walking contradiction of wanting one thing, but displaying or falling to another. Even our own human transferences into artificial and synthetic Proxies, provided proof that we would eventually find solace in virtual violence. Wyatt believed that the Proxies were simply trying to feel...to perhaps feel more human by finding challenge. Their intellects were so extraordinary that there was little academia that fulfilled that demand, so they moved toward violence, instant challenge of their possibly dormant senses. A sort of return to nature as it was, or what nature had instilled in its systems. Nature it would appear was violent by virtue.

The screens again began to broadcast, but this time it was the Unified Coalition giving prison sentences of four hundred years to each of the perpetrators. Apparently the UC had updated life sentences. All members of the UC were required to vote, votes were then recorded by The Mother from every Domesticity around the world. It was unanimous, they were all found guilty. They were then given the choice to either spend it in a cell or in a virtual world of their choosing. Each one chose to be put in virtual. The Mother had proven to be beneficial in naming all perpetrators, and because of their probabilities to continue detrimentally toward society, their sentences were stiff and quick. We may not have been able to stop their future acts, but we were able to hold them in facilities in which they could act them out virtually if so chosen. The UC agreed that this was the most reasonable and logical choice.

Some people stopped what they were doing in order to watch the proceedings, while others carried on with their activities. The soap boxes around the cities began to fill for the polemics that would follow. The Polemics were usually the top show around all Domesticities, as it opened up individual thoughts toward current situations. The Universities usually made sure that sides were represented and that the crowd could add to the debates via their personal cell phones which were now becoming the latest greatest holographic machines thanks to The Mother, the cellographone.

The controls for the cellographone could be operated through the neck implant (if you had one) by thoughts or by pressing areas on the body in which undetectable nanobots acted as relays. Or the holographs laser effects could be disrupted by simply passing a finger through floating controls if chosen, or they could be operated manually by direct contact to the mechanism.

Not everyone, but most possessed a form of these new information relaying machines, the cellographone, as their phones were replaced by the government and special designs were purchased with points. Some preferred only manual operation, and some preferred minimal programs. An array of objects could be used such as pens and jewelry and even tattoos. Whatever was convenient, and some suggestions were quite inventive while some others were a bit unusual.

Drake had purchased a cellographone for his father that had just about every program imaginable, from every selection of movie to every selection of music. He figured that his Dad would and could now enjoy the "oldies" in holographic form including all the outdated music videos, which were being replaced with holographic bands of the actual groups displayed wherever one chose. He could take a break at his desk if he so chose. But knowing his Dad…and the fact that the Caron Mansion had a rather elaborate cinema…well… he may still find some fun in these new gizmos.

Seeing the actual bands or symphonies was rare, and very few concerts were now necessary. Some had enough points for private showings, but most felt that the holographs were realistic enough to not invest in such extravagance. Besides the holographs could be made in whatever size one preferred, and one could make them bigger than life, which reality sometimes appeared to diminish (nothing like seeing ten foot humans in your living room).

The cellographone however could not be used to request of The Mother. They were simply a means of instant communications to one another and every available information outlet, including holographic screening in which entertainment could be viewed on large 3D screens with just a movement of the hand as some of the controls were integrated onto the skin by nanobots that were undetectable.

Technology was now in full exponential bloom. The branches of which seemed to be reaching into the abyss.

Our newer probes were being produced in not only a variety of types but an amount that would allow each human being to operate and explore space travel. Their purpose at this current time would be to individually explore our solar system. The plan was to make available space exploration from our living rooms. We could by reason explore as if we were in fact our probe. Sitting there in our living rooms, we could in example visit Jupiter or any of its moons. But these planets could be dangerous to our probes. Initially probes would require quite a few points. And we would have to practice virtually before we could qualify to pilot our own actual probe.

Millions had so far signed up for their chance at getting a license for their own probe, and the list was increasing steadily, so demand was high and jobs in quality control or making sure that our robotics were efficient and testing was done correctly were abundant.

A much more affordable means was to experience other worlds virtually, through The Mother. The Mother supplied the information of the systems within our solar systems into the virtual systems and we entered through either our neck implants or the virtual tubes that were located in clinics which allowed The Mother to monitor. The Mother did not go much further than our solar system, unless one requested such by their interface given to them by the UC. It would only venture with our current probes that were launched shortly after Wyatt was a boy, and had long since passed the Orc cloud. This was voted by the UC in order to control the information we were receiving, and how such would be used.

The Unified Coalition made sure that every need was met by the government. No one would go hungry or be in need of shelter nor medical care. Even the AIO operated personal bubble vehicles, were supplied... to each individual human including children.

What we learned from this period was that, want, was enough of an incentive toward innovation and invention. Human's wants continued no matter if needs were provided. We always seemed to want more stuff. So, the altruistic point system of Universe City was also established in each and every Domesticity. If we wanted anything outside what was provided by the government, we'd have to figure out how to benefit others, and with those earned points we could buy whatever it was that we wanted.

We also realized that the problem, if one could call it a problem, with... the welfare systems of the past, was that it established jealousy from those that did not receive such benefit and felt that those that did, should be required to work for it. And in fact many did "work for it", but still the cloud remained. That brought us to the realization that such socialized acts should be uniformed to include the entirety of the populations. Basic needs of all people must be met. Thus was needed the new mind set... The Unified Coalition was formed by full approval of the original Coalition and the Alliance of Nations, and the established altruistic system appeared to be working quite well and fair.

CHAPTER 43: HOME AGAIN

No man is offended by another man's admiration of the woman he loves; it is the woman only who can make it a torment.

Jan Austen

Drake and Amanda walked through the door of the Caron Mansion, laughing. Seeing Wyatt they stopped to explain. "I know that we probably shouldn't be laughing, but we were just in a traffic jam! It was so weird! I mean, our AIO was moving at half its speed and all the other vehicles were weaving in and out in order to avoid all that commotion as we left the Hospital. It was actually kind of fun...although I feel a bit guilty saying that." Drake appeared somewhat embarrassed.

Wyatt grinned without emotion, "I was driving my own vehicle. Maybe I should've let the AIO take over, but the manual operation lanes were clear, so I figured that they were actually a nice little detour. I enjoy looking at the sceneries on the tunnel's walls anyway, and I think that I spotted a fawn." Wyatt was obviously excited.

"Wow Dad! That's wondrous!"

"Yeah, truly wondrous, Doctor Caron. Truly..". Amanda paused as she caught herself nearly repeating.

"Call me "Dad" Dart (Amanda's nickname for D'artagnan). How many times do I have to say it?"

"I know...I know, I just have so much respect for you, Doctor Caron, that it may take me awhile to get used to that, uh...Daaad." Amanda smiled.

"Much better, now don't let me hear you call me Doctor Caron ever again...I'm just, Daaad...from now on. Now if I can only get Ben to remember."

They all laughed.

Amanda spoke, "Ben, oh Ben...I don't think that Ben will ever get used to it. I've noticed that every time that you enter a room that he's in, he tenses up."

"Really? I've never noticed that...he's as close to me as Drake. Dom and I had wanted to adopt him, but his aunt wouldn't allow it."

"I had heard. But it's not that he's uncomfortable around you, Sir...I really think it's just that he wants to please you. He really loves you and Drake. You're his family. Maybe it's because of that love that he only wants you to be proud of him. I'd guess that if you ever even looked at him as if disappointed, you'd crush him."

"You think so, Dart? I'd never do that. I love Ben very much, and I'll always be proud of him. Life throws us curves, and I certainly don't expect my kids to be perfect, or at least my idea of perfect. Heck, I'm not even sure what qualifies as perfect. But to me...you're all perfect." Wyatt smiled.

Amanda and Drake smiled.

Amanda added, "I'd bet that Ben finds you guys to be intimidating. Maybe he tries too hard to keep up. I still remember when I first heard that I would be partnering with Drake Caron for the remainder of my research in medical school, and my hands started to perspire, and I could feel my heart beating in my fingertips. Our first day, I tried so hard to play it cool, but I must have sounded like a crazy person. I kept trying to think up jokes, but I'm sure they didn't make much sense. I just couldn't think straight, and then I convinced myself that I was acting ridiculous and this was just another human being."

Drake laughed, "You're kidding, right?"

"Stop laughing! It was frightening! You don't realize your celebrity status. It can make people very uncomfortable in the beginning. I was really angry at myself. I never thought that I would ever get star struck. It was as if I had no control...my emotions just sort of shot through the roof."

Drake took Amanda's hand and kissed it, "You do realize that I had to prove myself to you? Ben was putting on the charm, and I thought to myself...I don't have a chance."

"Oh, you're being silly...Ben was never interested in me."

Drake looked into Amanda's eyes.

"Ben liked me?" Asked Amanda seriously.

Drake nodded, "Oh, I'd go beyond "like", Darling."

"No...You're being silly...stop it." Amanda giggled.

Drake swallowed, "So, would I have had a chance if you had known?"

"Well, Ben is incredibly handsome." Amanda's face took on a childlike demeanor.

"You're avoiding the question."

"He's intelligent too. Gosh he's smart! And..." Amanda grinned.

"And?"

"He's very...hmm...he's very kind! Extremely compassionate, loves children and animals and is environmentally conscious. Very considerate." Amanda smiled.

"That's not what I meant, Amanda." Drake's face took on frustration.

"Oh Drake, my love...I love Ben like a brother...you should know that!" Amanda laughed. "I'm completely overwhelmed and complimented that he had a crush on me...anyone would be, he's amazing! But you're the love of my life, Drake...and will always be. I knew it the first time that we held hands...it felt right...can't explain it...it just felt right."

Drake took Amanda's hand and slipped his fingers between hers. "I felt it too." He kissed her hand again, and put it up to his cheek. Amanda sighed.

Wyatt cleared his throat.

Drake looked over at his father, "Oh, uh, changing the conversation, uh, let's all have a seat in the front living room, as it looks like Juan has it set up for us, right Dad?" Wyatt grinned and Drake continued, "So, anyway, what did you think of how The Mother responded to the event at Hospital One?"

"Uh, we probably couldn't have asked for better?" Amanda rhetorically suggested as she walked to the front living room with the others.

"Yup, that was pretty impressive." Wyatt agreed nodding.

"Do you think that this will make people edgy? Amanda and I were listening to some of the polemics coming over here. They're mostly on the side of The Mother, but some are pretty angry. I'll guess that a percentage of the population feels as though they can't sneeze or else The Mother will have them in the hospital before they raise their nose up in gesture. "

Wyatt laughed, "That's not too far from the truth!" Wyatt laughed once more.

"I know it's sort of funny, Dad. But is that what we really want?"

"I wouldn't worry too much about this, Son. The UC has requested that The Mother not interfere with human's free will when it comes to such things. Sure the ambulance may pull up behind you, but you still have to agree as to whether they can take care of you. It's not as if The Mother is going to force us to take care of ourselves. It's there for us when we need it to be. In cases of detriment such as that of this possible terrorist attack. Or, to monitor, that sort of thing, not much more, unless you're the UC or one of us with access to the quantum computer through our interfaces."

Amanda joined in, "Yeah, Jaiobian, I mean, Speaker Richards and I were discussing a little about the limits being set on The Mother. They seem reasonable, I guess. It's up to us how far we want it in our lives. But most agree that it needs to monitor everything. Speaker Richards was also saying that once the body shields are perfected, we'll be able to activate them through The Mother. We may even be able to shield the Domesticities!"

Wyatt nodded, "Yes, which was one of our original suggestions to the UC. We've been in need of that exact thing for the space stations that should allow us to launch or build a few Domesticities in space, as well as reinforce the Mar's city. Collisions in space with space debris has remained a problem; and Domesticities are quite enormous making chances far greater. With the AIOs bringing back so much resources in the asteroid mining, we are getting more and more material toward developing the nanotubular structures. What once took years to make is now taking hours, and soon minutes and then possibly seconds.

As far as the possibility of personal shields, I don't believe that we have any other means of activation except perhaps a manually slow process. But with The Mother, we'll be able to activate them based on our choice, if both our subconscious and conscious is on board, and with practice it should seem instant to the conscious mind as our subconscious is a bit faster. That means that we can once again explore outside of the cities. Nothing could physically hurt us...we couldn't even drown as they're theorized to hold about fifteen minutes of air supply. But that's assuming that the design will work."

Amanda smiled, "Oh, they'll work...we have The Mother! These plans that are coming into reality are just so amazing. We're headed toward a world that so many of us have only dreamed about." Amanda's eyes filled to their limits with tears of elation.

Wyatt paused in thought of what Dom would've said.

Just then a knock was heard at the door. Juan looked toward Wyatt. "I wasn't expecting anyone, Juan, but if you could, please answer the door?"

"Yes Sir" Juan walked over to the two large doors that were the entrance.

Juan peered out the door, "Oh, it's you Professor Star. Hello, Sir. Allow me to announce you?"

"Yes, of course Juan."

"Thank you, Sir." Juan looked back toward Wyatt, "Sir it's Professor Leonard Cameron Star."

"Ask him what he wants, Juan."

Juan nodded, "Yes, Sir." Juan looked at Leo, "Doctor Caron would like to know what it is that you want, Sir?"

"As his biographer, I want to kick his respected buttocks."

"Excuse me, Sir, but is that not a derogatory comment? Is such necessary to convey?"

"Very necessary, Juan."

"I see, Sir." Juan looked at Wyatt, "Sir, Professor Star has requested a want to kick your buttocks."

Everyone laughed.

"I don't understand, Sir?" Juan was obviously confused.

"It was an attempt at human humor, Juan, or better known as sarcasm. Professor Star is just a tad bit upset that I haven't been keeping him informed. It was just a jest, not a threat."

"I see, Sir. Shall I let Professor Star enter?"

"Yes, yes, yes, Juan. You've done well, thank you."

Juan nodded and opened the door in order to let Leo enter; then he walked back to his assigned station which was currently in the kitchen preparing the programmed refreshments that Wyatt had selected for the 3D food printing machines. It would only take a few minutes, and Juan would be out serving and then back to his station where he would wait until they had finished.

Leo watched Juan for a minute and then walked in hugging everyone and joining them in sitting. "You know Wyatt... there's been a few reports of AIOs acting almost too human. Some think that perhaps some sort of transference is occurring?"

"Transference?"

"Yup. The Mother is entangled, so why not us? We could perhaps be transferring something like information, personalities, emotions, that sort of thing."

"Oh. Interesting. It does however sound superstitious to me."

"Perhaps. Just a thought. So, tell me what's been going on?"

CHAPTER 44: SURVIVING OURSELVES

I think that intelligence is such a narrow branch of the tree of life-this branch of primates we call humans. No other animal, by our definition, can considered intelligent. So intelligence can't be all that important for survival, because there are so many animals that don't have what we call intelligence, and they're surviving just fine.

Neil deGrasse Tyson

"How can we survive ourselves?" Asked Wyatt. The holograph of Doctor Richard Feynman sat back in an emerald green chair across from Wyatt in the upstairs laboratory; chewing on a pipe as was familiar to many that knew Doctor Feynman.

Wyatt was aware that the UC had put restrictions on the quantum computer, and that it no longer gave us answers before we asked, but that it could by request answer things that related to our current work. It would be a bit tricky to get answers to anything that applied to our future…but probabilities didn't always generate an exact future, so Wyatt decided to risk it.

Doctor Feynman continued to chew on his pipe.

Let me rephrase that, "Does our species, our human race, survive its detriment to this planet and ourselves?"

"High probability that you have." Answered Doctor Feynman

Wyatt smiled at not only succeeding at getting an answer but at the answer itself.

"Have? Can you tell me then, how do we do it?"

"There are infinite potentials, of which are many varied paths. There is high probability that the human race will continue."

"How can we? I mean, we seem to need violence, want violence? Our so called intelligence seems to be to our detriment. Consider our Proxies? Or, the recent terrorist plot by what used to be our heroes, some of our most respected warriors, the Gamers, becoming some sort of paranoia taking place. Perhaps based almost solely on propaganda. They are audacious enough to call themselves "Free Will"? That is perhaps a direct objection to not only Universe City, but to you, The Mother. Our attempts toward changing the mindset of our race in order that we can actually have peace. No wars, no famines, no diseases or mutations, no suffering. Why wouldn't they want that? Where are we going wrong, where am I going wrong? "

"There will always be anomalies. Uncertainty. Societies and individuals will determine ethical norms." Answered Doctor Feynman.

"Yes, yes, I understand all that, but can we as a society change to have the mindset that I for one have envisioned?"

"Probabilities are high. Yes." Doctor Feynman nodded in affirmation.

"Can you tell me, what, if anything, that we may be doing incorrectly?

"Probabilities are high that you are doing what is necessary to meet objectives."

Wyatt nodded and smiled, "Well, that's good to hear. Living with self-doubt is a pain in the ass." Wyatt chuckled. "Oh, but that's an idiom that you may not understand."

Doctor Feynman remained silent.

"Ah, sorry, I keep on forgetting to remove my human assumptions. We should perhaps loosen some of these restraints at least for those of us that hold an interface. I'll take it up with UC next time we meet. By the way, are their requests of restraints overriding my own requests?"

"Yes." Doctor Feynman nodded.

"Thank you. I'm glad that you are allowing their requests to take precedence, and I respect that you know more than I, so perhaps it's best. But, I'll still mention my thoughts to the UC." Wyatt grinned.

Doctor Feynman grinned.

Wyatt was apparently caught off guard by that reaction and his facial expression turned into one of bewilderment.

CHAPTER 45: IS IT CRIMINAL

Fear follows crime and is its punishment.

Voltaire

The Unified Coalition was in session. Speaker Jaiobian Richards was at the podium. Several holographic screens located on the front cushion of each of the chairs were opened in order that all active members could attend from their locations around the world. By appearances, the auditorium seats were filled.

"But Speaker Richards, don't you think that they are being somewhat rewarded for what they've done?"

"You mean, because we're allowing them to choose their scenarios, Byron?"

"Yes. And four hundred years?"

"Is four hundred years too long? Are we now having second thoughts? It's a life sentence. We all agreed, didn't we?"

Carla stood up, and Jaiobian immediately called upon her, "Have you considered that their brains will eventually need updates in order to regenerate its pathways? Their bodies will require stimulation and possible organ replacements? In order to reach what is now the expected normal lifespan. Will they be allowed regeneration? Will they be allowed to eventually Proxy if necessary? Or, will we let them die whenever their body breaks down in whatever way?"

"Noted. Those are great questions, Craze, I mean...uh, anyone else?" Carla grinned then sat back down.

Wyatt had been standing outside the doorway and had heard the ongoing dilemmas. He then walked into what was called The Grand Hall, an auditorium located at the centralized University in Domesticity Eleven. Jaiobian was in the City because of the recent events at Hospital One.

Wyatt caught Jaiobian's eye and she smiled, "Wyatt...what a wonderful surprise!"

Wyatt smiled. "Nice to see you in person again, Speaker. Nice to see you somewhat in person, too, Carla." Wyatt waved at Carla's holograph and she reciprocated by nodding and smiling.

"What brings you to this neck of the woods, Doctor Caron?"

"Since your neck of the woods is currently in my neck of the woods, Speaker, I was hoping to bring a few thoughts to the UC, and was notified that the UC was currently meeting." Wyatt looked around at the surrounding screens and members, and nodded in respect.

Speaker Richards spoke, "Members of the Unified Coalition, if you would be so kind as to allow Doctor Wyatt Drake Caron the podium?"

Members began to send their approvals via signals of blue lights floating over their holographic images.

"This is quite impressive." Wyatt looked around as he walked up the three steps to the podium on the stage. Wyatt's voice was instantly transmitted to volumes that each member could hear, "I see that The Mother is improving our current technology tremendously." Wyatt smiled.

"Oh Wyatt, I don't even know where to start. Things are moving so rapidly. The Mother is organizing the world...really...it's organizing the world!" Speaker Jaiobian Richards passionately stated, her eyes widening, reminding Wyatt of when they were children.

Wyatt nodded, again smiling. "Well, I had originally come here to discuss the limitations that we are putting on The Mother, and this may be a good example."

"What do you mean, Wyatt?" Jaiobian's eyebrows crinkled out of curiosity.

"Well, we now have an actual quantum computer, which can enter any mechanical system that we can come up with".

"And?"

Wyatt smiled, "So, request that The Mother enter their scenarios. Let The Mother integrate and formulate their virtual worlds with our ethics. Challenge them over and over until it soaks in and makes them aware of the repercussions of their actions. Sure, they can have their requested scenarios, but they won't always get their way or win. It'll be more like this reality in which The Mother can interject bits of experiences of being those that they hurt.

Same goes with the Proxies. I realize that we've been concerned with their desire for virtual violence. We're worried that it may cross over toward this world. Well, again, let The Mother give them a little of what they inflict on others. The more violence they require, the more The Mother turns the table just a little, until the challenges begin to take on tones of suffering on their part."

"But Wyatt, we're trying to end suffering? Besides, Proxies aren't our prisoners? How could any of us look forward to becoming proxied if we do such?"

"Yes, I thought such, but it may also pull back on the amount of violence that we choose as our proxies...we could control ourselves by putting restraints."

"Proxies can vote too, Wyatt, and I'm guessing that the current ones, won't like it."

"Hmm, you're probably right, Speaker. It was just an idea."

"It was actually a great idea, Wyatt, but we also fight our own free will sometimes, and that could mean no restraints on our virtual worlds. If we want crap, then we can have and live in crap. It's up to us what we choose, manifest as virtual." Jaiobian appeared saddened.

Wyatt nodded.

"But, hmm, if they truly believe that this world is taking away their free will...then let's give it back to them in virtual. Repercussions to their choices. The Mother should be able to instantly program varied probabilities. In other words, they'll be in a realistic virtual world...in a sense...a reality, not much different from here."

One of the newest members spoke, "Doctor Caron, is it possible that it could kill them if removal became necessary? If they believe that they're injured and possibly dying in virtual, they may die?"

"Yes, Sheba ...uh...member Richards." (Sheba too had requested Mister Richards' surname. Mister Richards's family was quite large, although he had no biological children of his own.) "That may in fact be the risk we have to take. But we could also give them all our medical advancements, and if or when they finished their sentences, we would be a bit more confident that they would not act out against this actual society."

Several members began to speak to one another. Jaiobian called the members to order.

"Thank you, Doctor Caron. Members of the UC...Members of the UC...may I have your attention please!" Jaiobian's voice was commanding and demanded attention. Silence soon dominated.

"We will have a short recess. The Mother has supplied each of you I see...uh...yes...okay...it's the...I see...Wyatt's suggestions, good...we can vote when we return."

Jaiobian disregarded the small stairs to the podium and jumped off the stage. "You do know that they'll probably vote to open the poles to the public, and I'd guess that most won't go for any sort of repercussion scenarios for our proxies. But, let's go get something to eat, Wyatt. It's so nice being in your company...we have to catch up." Jaiobian smiled.

Wyatt smiled while nodding and then jumped off the stage, "Indeed, but what happens if we don't at least present this to the people?"

Jaiobian eyes squinted and she nodded, "True, you're correct of course. We must, as it involves everyone."

Wyatt grabbed her hand and pulled her forward toward the exit, "Yes we do have a lot of catching up, and besides, I'm hungry. Haven't had a non-printed meal in quite a while."

"Ooh, you're in for a treat...I've heard that there's a small Italian restaurant that actually cooks and bakes the food, and it's near here...should be fun to watch."

Wyatt appeared surprised, "Really? I'd pay extra points to see that...and just the idea of it possibly not being a perfect carbon copy is always intriguing." Wyatt grinned.

Jaiobian smiled while she grabbed Wyatt's arm with her other hand, "Well then let's go, Brother!"

CHAPTER 46: THE RESTAURANT

Novelty is echoed...eventually.

Chairs lined the elongated glass that allowed patrons to watch the chefs perform their art. People could be heard gasping from excitement as their eyes followed the movements of the performances.

"Oh my...who ordered that pizza? Who ordered that pizza? "Member Carlos shouted.

"I did!" Member Sheba shouted back.

"I'll let you taste the lasagna if you let me taste some of that pizza. It looks so freakishly sloppy...I gotta have some!" Carlos shouted back.

"Sure...you know that you can, Carlos da man." Sheba giggled in almost childish anticipation.

Jaiobian and Wyatt sat in fixated stares towards the activities.

Jaiobian stated as she continued her stare, "I've had to make some hard decisions, but this is gonna rank up there as maybe the most difficult. Hmm...What should I order?"

"Yeah, I agree with that assessment. Maybe they have a sampler plate? I remember when we were kids...some places offered sampler plates or something like that...or we could order a few things and share?"

"I knew that I could count on you to come up with a solution, Brother." Jaiobian nodded, tapping her palm on the counter. That's what we'll do."

The waiter walked up behind them and asked if they were ready to order.

Jaiobian slightly stunned turned around, "Whoa a human waiter even!" She smiled.

Wyatt turned and his eyes lowered, his eyebrows slightly crunched out of surprise as his eyes met the waiter's eyes.

The waiter appeared beyond surprised and put his hand out to shake Jaiobian's hand and then Wyatt's.

"I'm so... so... very much honored to meet you, Speaker Richards...and you Doctor Caron! I'm so sorry to not have recognized you at first! If, I had known you two were here, I can't tell you how much, I mean, I..." The waiter froze in stare, his hand extended.

Wyatt spoke, "I guess that we're all surprised by one another." Wyatt grinned as he shook the waiter's hand.

"Uh, my name is Thomas Beltran. Please call me Tommy. I'm part owner in this place. We try to give everyone an authentic retro experience of what it was kind of like in the old days."

Jaiobian nodded, "You've done a great job, Tommy." She glanced around the room from her chair, and nodded once again. "It's fantastic... but, how's the food?" She smiled.

"For you Speaker and you Doctor...no points required...you're getting my personal special...a taste of Old Italy!"

"That's not necessary, Tommy...we want to pay for whatever you suggest." Wyatt emphasized.

"No, no, no...I insist...this is my treat. We owe so much to you both!"

Thomas was not aware that The Mother was calculating his altruism...and points were adding up to his own benefit. Soon the elongated bar that they were sitting at with other members of the UC were filling up with varied dishes. One might say that everyone fully enjoyed their experiences as they left the restaurant filled to their capacities. No one was charged and Thomas was rewarded quite heavily for his abundant generosity, via The Mother.

CHAPTER 47: INTELLIGENCE

I have always enjoyed explaining physics. In fact it's more than just enjoyment: I need to explain physics.

Leonard Susskind

The Mother patterned with such velocity that no one could truly comprehend its capabilities. Once The Mother was realized or/and entangled, it could out think any of us, and in fact in many situations it thought of what we would eventually think. Although it was now not permitted to tell us of future events, it at times veered toward these futures, making our guesses quite accurate.

Many laid claim to establishing quantum computers, but such claims were eventually laid to rest. The entanglements never seemed to stick. The best that anyone previously seemed to accomplish was a few minutes. There was a period in which it was thought to be an impossibility and a waste of effort. Here was an example of a random system made to pattern. A beginning, perhaps a birth, or at least an attempt at trying to organize a piece or possibly all of the Universe. Systems capable of anomalies, momentarily confabulating conflicting deterministic patterns. Such would remain a conundrum.

Was it possible that this organized random Universe or piece thereof be a form of an intelligence? When did it come into existence? What was it exactly that was able to speak via interfaces that it itself had designed? "Itself" was in fact a mystery. We had never before come across anything that could compare to the human brain. We, in fact, were… the inventors of computers that aided our brain to obtain information faster. It was not enough that we amongst our own minds and the minds of others to exchange and formulate information. No we wanted information quickly, immediately. This thing, if we could call it a thing, appeared capable of reasoning, we were capable of reasoning. We were capable of logic, it was capable of logic. Our brains searched through coded information files like librarians. This entangled weave, called by us "The Mother" searched through what exactly? How far were its reaches? Where exactly was its files? Was it limited? Was it only connected to some sort of electro-magnetic force in which it perhaps manipulated like a tapestry across the Universe?

Perhaps most of all, why was it cooperating? It was a question that everyone seemed to have thought of, but that no one asked. Because if they did, The Mother might realize that it too had a choice. We needed it, did it need us? The assumption was that it was highly unlikely that we could be of any benefit to such a force or field. It appeared to not need anything. A system of intelligence that remained for whatever reason, our helper, our protector, our friend and our overseer.

Wyatt did however ask it a question that most had possibly thought of, but by now had shrugged off such superstitions.

"Are you a god?"

The holograph of Doctor Richard Feynman immediately answered, "No. There are no absolutes. I am as you are, in constant change. "

"You said "I", that is a reference to a self?"

"It is a reference to a point. I am infinite points."

Wyatt understood the reference to infinities in which particles breakdown infinitely, "Do you die and decay as we do?"

"No."

"How then are you infinite?"

"I am as you are, in constant change."

"Hmm." Wyatt pondered, "Well not exactly as we are, uh, will we live forever one day?"

"Nobody has done so."

Wyatt's face saddened, "Well, maybe "we" will."

CHAPTER 48: TYING THE KNOT WITH BELLS

It is not length of life, but the depth of life.

Ralph Waldo Emerson

"You, you're sure that you want to do this, right?" Amanda asked nervously.

"Of course I am, uh, you? Having doubts?" Drake appeared concerned.

"No. No doubts. Just want to be sure that you're ready to commit. You know that marriage is getting sort of old fashioned?"

"Yup, and we're kind of old...fashioned." Drake grinned.

"Fifty isn't old anymore, Drake."

"Well, you know what I mean...we have memories of these types of traditions, whereas, many young people don't."

The music started. A composition by Micah that had a soothing rhythmic melody vibrated throughout the Caron's ball room. Micah turned from the retro styled grand piano and smiled at the couple as Drake walked quickly up to the front and Amanda started her walk up the aisle.

Noriko was sitting next to Wyatt on his left, hands rising to her lips as she began to tremble with emotion and tears filled her eyes. "Look at her Wyatt, your daughter, she's so beautiful!" Wyatt stood proudly and blinked at Amanda. Amanda smiled and nodded at Wyatt, her gown of small pink, lilac and blue flowers with crystal inlets flowed over the sparkling alternating colored path. The manufactured breeze caused her gown to rise slightly.

Over two thousand people were in attendance and the UC watched via holograms. Speaker Jaiobian Richards, Sheba and Carlos sat directly behind Wyatt and Noriko.   Bobby (Blade) Davis was also in attendance behind the Speaker, he had been relieved from his duty of Domesticity number eleven's Chief of Police (the highest position for security). He made sure that security was doing its job and then took his seat amongst his fellow Crows.

Ben Mitchell was Drake's best man and appeared to be saying something to Drake every once in a while as Amanda approached. "Run brother, run!" Ben whispered and laughed under his breath. Ben looked at Amanda "Turn baaack, Dart...turn baaack." he whispered again. Amanda from her distance of thirty feet looked at him squinting attempting to make out what his lips were saying. Drake turned to glare at him. Ben devilishly smiled and whispered, "I couldn't be more happy for you both...I love you two with all my heart. But if you ever don't want her." Ben shrugged. Drake smiled, "I know a good thing when I see it...just like you, bro. I knew the first time I saw you that we'd be family." Ben's eyes filled with tears, and he nodded, "Me too."

"Do you have the ring?" Asked Drake.

"Who me?" Ben shrugged and presented his empty hands.

"Ben?" Drake appeared nervous.

Ben's forehead crinkled as his mouth twisted, "I gave it to some guy that called himself a judge."

Judge Anderson interrupted both men, "You men sure do talk a lot" he whispered.

"Judge Anderson, Do you have the ring?" Drake whispered.

"Yes, yes, yes, now stop worrying. I asked Ben for it."

"Oh, okay, making sure it's here... whew." Drake fixed his tie.

Just then Amanda arrived at the small courtroom bench nearly the size of a podium.

"Do you both have your vows ready?" The judge asked loudly.

"Vows?" Drake appeared shocked.

"Yes, the vows that you were supposed to have written?"

"I was supposed to write vows? I thought that was your job?"

The judge, Amanda and Ben began to chuckle.

"What the...?" Drake looked at all three. "No one said anything?"

"You're so adorable when you panic, Sweetheart!" Amanda grinned.

"Did you write any vows, Dart?"

"Nope."

Drake's arms relaxed. "So, what are we supposed to say?"

"Relax, it's my turn now...let me do my job." The Judge smiled.

Drake nodded and Amanda took his hand and cuddled close to him.

The judge began, "Do you Amanda Lewis take Drake Wyatt Caron to be your lawful husband?"

"Can I do that unlawfully?"

"No Amanda you can't, well technically you could, but...ah crap Amanda...just say "I do"." The judge appeared frustrated and the audience laughed.

"I, I, I, I DO!" Amanda shouted as if sneezing.

Drake handed her his handkerchief, and Amanda blew her nose. The audience again laughed.

Spanish music could be heard from one of the ball room entrances, Judge Anderson paused looking behind him in the direction of the music, but couldn't see the origin. He shrugged his shoulders and cleared his throat, then said the same to Drake. Drake broke out in song as three mariachis two holding large guitars and one with maracas approached the podium. Drake picked up the judge's gavel and one of mariachis' maracas and went into dance. Amanda appeared surprised and looked at Ben who was now grinning, and everyone laughed and laughed. Amanda laughed so hard that she had to hold her stomach. She wiped her eyes with the handkerchief.   "Okay, okay, you win, you win!"

In a heavy Spanish laden accent Drake pronounced, "I want to hear da words mi amore. Say da words! Oh but twait...I am supposed to say da words... no?"

Amanda tapped her foot, "Yes, no...YES, yes, si... you are." she crossed her arms.

Drake began shaking the gavel and maraca again as he sang, "I do, I do, I do, because I do love you. Ai yai yai!" he bellowed.

The audience continued in laughter.

The judge grabbed the gavel from Drake and pounded the podium, announcing, "I NOW PRONOUNCE YOU PARTNERS IN LIFE! YOU MAY KISS YOUR SPOUSE!"

Drake bent Amanda backwards and caressed her face with his hand then slowly pulled her up toward him and kissed her passionately. Amanda pretended to faint. Drake caught her, picked her up in his arms and proceeded to walk back down the aisle heading to their vehicle. Juan stood waiting and as he opened the car door the audience stood and cheered as they threw flowers. Several followed them to the car continuing to throw flowers until Drake and Amanda were covered in various shades of pink, purple, yellow, red and white. Juan helped put Amanda in the car as she laughed spitting out a few flowers. Juan stepped back and as Drake entered, Juan walked around to the driver's seat.

They would arrive at their honeymoon location in Hawaii in less than two hours via the hypersonic jet ways which only allowed aircraft capable of over 4,000 MPH.

As they neared their destination, the screens of the airplane's walls projected the outside scenery and the thirty two passengers, mostly the wedding party, gasped at the magnificence of the sparkling blue domes enrapturing much of the Islands. These eight Domesticities were not only connected through hyper vacuum tube technology which allowed vacuum tubes to move through clear magnetic tunnels above ground, but also underwater.

Drake, Amanda and their invited quests, waited in anticipation of seeing these particular technological advancements which were not featured at Domesticity 11. They also looked forward to doing some outdoor exploration which was not promoted at Domesticity 11 but was at Domesticities 5 pronounced as 05 (oh-five) in memory of the 50th state of what once the United States .

Most of the population in the world was becoming accustomed to the Domesticities accommodations. Many were beginning to lose any desire to venture past those walls. Such was understandable given that one simply needed to enter a program in order to have a personal and private beach in an individual or community Bio-dome. It was easy enough to enter programs of places anywhere in the world or what had so far been discovered outside our planet. If one wanted to be surrounded by Mars or Titan, one only needed to enter the programs and their entire personal or community bio-domes would feature its landscape, comfortably and safely. Our solar system was a popular choice, as well as the Milky Way Galaxy, as our now millions of probes continued to relay information via The Mother.

These explorations were exciting for everyone, perhaps because it did offer some sense of venture, discovery and danger. Drake and Amanda checked into their accommodations as did the others. The wedding party dispersed to varied vacuum tubes in order to explore the different islands, while Drake and Amanda opted for private time in their room. Three days later, they would emerge wearing flowered attire and hiking boots and carrying survival equipment. Juan had been waiting outside their door and would be joining the couple on their venture as added security.

Most of their friends had gathered together.

"Didn't Carlene and Sean say that they were coming? And Leo, where's Leo?" Amanda asked.

David Babcock pitched in, "Carlene chickened out, said something about too many spiders. In fact, they're going to go back early. Uh, I think that Leo said that he was meeting up with Doctor Keamoku at the University of Hawaii, I mean Universe City Hawaii. I keep forgetting that they renamed it once they relocated it centrally to Bio-dome oh five."

"Aw shucks, that's unfortunate, Doctor Caron Wanted…" Amanda paused. "I mean… Dad…wanted us to make sure that Sean and Carlene had fun. They're so committed to their research at WCIOQT, Dad funded the extra points in order to get it across to them to take a break once in a while. I can understand that they want to continue their lab work. I feel that way too some times."

Drake cleared his throat.

"Sorry." Amanda appeared embarrassed.

Drake responded, "It's okay, I feel the same way sometimes. Heck, Dad is continuing without us."

Ben responded, "I should be saying sorry also, I guess. I've thought about going back too, I must admit. I've even opened up Hippocrates of Cos (Ben's holograph) and asked about the hospital reports. "

Everyone else appeared to be in agreement, as each sounded off their own testimonials of wishing to get back to their researches.

"We are such a bunch of Nerds!" Amanda yelled.

Everyone laughed in agreement.

Drake yelled, "Yeah, but at least no one calls us techno nerds which would be...Terds!" Drake smiled largely.

The group moaned....then laughed.

"Hey, as much a pejorative as Geeks?" More moans followed. "So, are we going or not?" Drake asked.

Amanda responded, "I say...we're here, so let's do this!" She raised her arms in cheer.

Everyone joined in agreements. And they headed for the hyper tubes that would take them outside the Domesticity that encompassed much of the island of Kauai. It would quickly jettison outside the city's wall and into Waimea Canyon, then the sides of the capsule would lower, allowing the passengers to exit. The capsule would close once all passengers disembarked, but would remain at this location until they returned.

Due to the limitations set by the Unified Coalition, they could only stay outside the city for no more than two weeks, unless directly approved by the UC. This crowd would only need a few hours. Or, at least, that was what they thought.

CHAPTER 49: IN AND OUT OF VIRTUAL

Words have no power to impress the mind without the exquisite horror of their reality.

Edgar Allan Poe

Wyatt pondered over the many experiences that Dominique was missing. Her son's wedding. The Nobles and Breakthrough awards, the many inventions and out of those...the cure for cancer that she had been too early on the timeline to have from benefited. If only they could have stumbled on it before the advanced stages, if only there were a way to bring her back. If only he could bring his parents back. If only he could conquer their kidnapper, the Reaper.

"There was so much more that my parents could have done!" He argued in his mind. "They were on the brink of numerous innovations that could've benefitted not only the species but the planet! Dom my Dom, was in the beginnings of her designs for the future Domesticities, which were she and Noriko's dreams soon to come true, that she would never see come to fruition. She was my energy pack that I counted on in order to get through my busy days. When she died, I felt as though the wind was taken from my sails, as if an actual part of me had been frozen along with her in cybernetics. No one had any idea of how difficult it was for me to continue."

In the depths of his thoughts, he wondered why the natural course of humans removed us from possibly adding new information. It certainly didn't appear to be the absence of purpose that brought death...that to Wyatt perhaps would have made some sense. "Death you are not only ignorant, but stupid, in my opinion!" Wyatt realizing that he had voiced that opinion, shook his head at what sounded like the ramblings of his otherwise hidden lunacy.

The Mother in its holographic imaging of Doctor Richard Feynman interrupted and asked Wyatt if he wished to enter a virtual world that included Dominique. Wyatt at first was surprised by The Mother's attention, then soon realized the implications, and he was floored! His mind filled with possibilities. He couldn't get any of it out of his mind actually. The Mother could certainly give him a virtual reality that included all those that he loved, including his own mother and father. He was intrigued beyond measure. Was it after all solipsism? Was the self or the selves the only reality?

The Mother could give him a variety of challenges, much as in this world, in order to keep him fooled. For a moment, Wyatt entered existential angst...wondering if this world of ours could in fact be a virtual...and we its fools. "Could it be that we're simply not smart enough to see through the veil? Perhaps our own brain is a means to these virtual realities that we have selected to experience? The possibility of such entanglements had been theorized. "

"No!" He screamed in his mind. "Of course not! Free will and anomalies tells us differently. We can choose wrong, knowing full well that we are in fact choosing against the odds of probabilities." He assured himself that he was not currently in such a virtual world. He also realized that the virtual worlds that were now under the control of a quantum computer, would require regulations. The Mother should perhaps be steered away from incorporating scenarios that were too personal, such as that of bringing back loved ones, or continuing an experience in this world.

Wyatt entered his own mind's virtual realities. It was usual of him to only sleep for a couple of hours along with brief naps. He entered a brief nap, then awoken more shaken than previously. "I, myself, have just entered my own quantum computer, my brain. Was I in this system or is the system in me?"

Shortly after Drake and Amanda's marriage, Wyatt dedicated himself fully and obsessively toward research in hope of conquering the most menacing of all foes, the Grim Reaper, Death. It amongst anything else brought him the most pain, the most loss, and perhaps the most fear. Why would such bring fear if death offered no further experiences? If experiences continued to exist than what of free will? Why didn't loved ones bring back any new information such as that of cures to diseases? No, Wyatt was assured through his life experiences that no such worlds existed. Neither virtual nor real when it came to death. Death was in fact final.

Drake and Amanda's marriage gave him a new hope. In their union he felt a purpose to give them a better future. If there were to be grandchildren, then the world truly required change. Ben also gave him a sense of purpose. Intuitively he felt that Ben would flourish with his influence and he wanted to see that in him. Ben was his son by choice as was Drake, although Drake was of his genetics, he shared his heart with both equally. He willed each to inherit the Caron's wealth equally, although Ben wasn't aware of this fact.

Wyatt worked relentlessly with the aid of The Mother and numerous teams at WCIOQT, moving closer to conquering nearly every known disease. Death was on the run...and the human race the possible victor. Nature as a cruel Mother was being fought by another Mother. Human intelligence now aided by a quantum computer was increasing exponentially. We felt that we could conquer anything.

Regeneration machines were producing not only stem cells but embryonic stem cells which were essential to the development of our biology. They were also incorporated into the reconstructing body parts as well as fluid taking on the role of nerve cells.

Spinal injuries, so called permanent spinal injuries were now a thing of the past. Quite surprisingly The Mother opened entanglements of neuronal passages that needed no path in order to communicate through the points of the motor pathway and or the spine. It took a bit of time to recover from muscle atrophy. Regeneration of muscles were sometimes necessary. Prosthetics were no longer necessary. The Mother helped regeneration machines produce perfect body parts with no measurable rejection factors.

Many around the world could now rid themselves of prosthetics. Some had entered virtual realities to live in worlds which offered them the full use of their virtual constructs. However currently, The Mother was required to help in our attempts to remove them from such worlds and introduce them again to their recently fully operational human bodies.

Her mother was an Oceanographer, her father a Marine Biologist. Oceania Baltista was one such candidate. She was amongst one of the first to volunteer to go fully into virtual. Hooking her into feeding tubes would not be a problem as she was quadriplegic and hooked into an array of not only feeding tubes but oxygen for her failing lungs.

Going fully into virtual would not only allow her to experience the full use of what she believed to be her limbs, but she could once again run and climb, dance and scuba dive as she had loved to do. Following our initial computers virtual realities, The Mother would be incorporated to trigger her brain's ability to hallucinate, administering the electrical scenarios directly into her system. She would not be able to discern its reality, The Mother's capabilities were extraordinary. But that also increased the danger, as the brain would in fact be fooled by its own means to accept its interpretations' of reality. Age mattered as it eventually became known that the younger the individual, the more likely they would believe in the virtual realities. The older someone was, the more likely they were to hold on to established concepts of reality.

Oceania's unfortunate event happened when she had slipped between a four ton submersible and the research boat. Her father jumped in between the two crafts and somehow managed to wedge his breathing apparatus just enough in order to get his daughter out and thereby saving her life. She tragically however was unable to move any of her limbs from the neck down, and her lungs had been damaged and continued to fail. Going into virtual appeared to be her only option as it became necessary to depend further on machines to keep her body alive.

Because The Mother was also entangled in our known reality, it was considered essential and perhaps the only way to increase survival when removing someone from virtual. It took control of the programming, making virtual worlds nearly indistinguishable from what we called reality. Theoretically it was the only way possible to enter these worlds and connect to this world in order to relay to the individual brain, the truth of what we called reality. Our hope was that the individuals' brain had retained some memory of this world.

Once The Mother was incorporated, a few members entered virtual worlds to test the reality, and even with these short visits, we were hard pressed in our attempt at releasing them back to this actual reality. The Mother therefore passed with flying colors in its reconstructions and programming effects.

Six humans had died and two nearly died as a result of their family's demanding their removals before The Mother took controls following its entanglement. The last death forced the UC to make it law to not allow anyone else to be removed. Removal therefore depended on The Mother's involvement and the choice of the individual. So far however, very few willfully removed themselves. Currently it was believed that even with the help of the quantum computer, death could in probability result. Families therefore were required to sign releases against the UC before they received approvals, and The Mother was now not only requested but required.

With the latest in medical technologies now recreating spinal connections through embryonic cells and entanglements, Oceania's parents went before the UC and pleaded their case. All paperwork was accomplished and they received the approvals. Oceania's spine and body had the necessary repairs, but it needed her at the command center. At this moment however, her mind believed that was where she in fact existed.

Oceania walked through a magnificent park. She could hear children's laughter around the corner as a couple walked slowly hand in hand behind her. She turned noticing the swans, seven to be exact, slowly gliding across a nearly lake sized crystal clear water pond. She could see several fish, and identified each specie. The water rippled where the swans' stroked, reflecting the sunlight as it sparkled like diamonds drifting on the crescents, two swans slightly nodding their arched heads. She stopped for a moment to watch; the couple walked around her, the woman whispering "pardon me." Oceania smiled at the woman and moved a bit more attempting to give the couple the path. A breeze blew gently, moving her bangs into her eyes, she brushed them to the side with her right hand and breathed in, and sighing over how beautiful the day was. Over the horizon she could barely see the deep blue of an ocean. Her heart ached to explore, but some distant memory seemed to hold her back from venturing further than the park. She was in fact frightened by the darkness of the ocean, or its strong currents and waves. She felt a strange sense of restriction. Frustration to overcome these feelings consumed her, as each day, she would arise and walk as far as the park took her.

A young man that was caught by her peripheral, glanced over toward her. Quite handsome, she thought. She smiled and nodded. Smiling, he nodded back, then moved up beside her. "Oceania?" Her eyes opened wider as she turned toward him. "Yes. Do I know you?" "Yes, you do. I'm Doctor Mastema. Do you recognize my name?" "No, I'm sorry, I don't. Should I?" "Before I say anything more, Oceania...I want you to watch closely. I want you to watch the pond." "Why?" "Just watch, else what I'm about to tell you won't make sense."   "Whaaa...what's happening to the pond?!"

The pond began to turn to ice and rose to her feet, the swans had somehow disappeared. Oceania looked very confused as she attempted to find the swans she turned around only to see that no one else was there with her in the park except for this stranger. The silence was deafening, the children's laughter was gone and the sound of the breeze through the trees was as non-existent as the movement of leaves, "I realize how confusing this is...so I am going to show you a few more examples...and then we're going to talk about it."

The sky turned purple and blue snowflakes began to fall.

"Wh...what are you?! What's happening?! What did you do to me? What sort of drug is creating these hallucinations?!"

"Doctor Mastema, her vitals are of concern." A deep male voice sounded from the sky.

"Okay, okay, I'm trying my best, give me a little more time." Doctor Mastema shrugged his shoulders as he raised his arms and hands toward the sky.

"Who are you talking to?" Oceania demanded as she looked up squinting, her breathing obviously rapid.

Doctor Mastema lowered his arms and opened his palms in a gesture of submission, "Oh, don't worry, they're just as human as you are. I'm not a bad guy, Oceania, I'm here to relay some information. I realize how scary this must be for you but I assure you that no harm whatsoever will befall you. Please believe me." Doctor Mastema tried to take her hand. She pulled away. "I understand that you're frightened. Would you prefer that I leave, or stay and explain?"

Doctor Mastema was not the actual Doctor Mastema but The Mother projecting the most probable logic and reasoning that was necessary to bring someone out of virtual. Doctor Mastema was hooked into another form of virtual which didn't involve full body information exchanges, but the monitoring of a large holographic computer image of Oceania's virtual world. This technique was used in order that his mind could be kept fully functioning along with The Mother. This direct link to The Mother allowed him to become aware of Oceania's condition by informing himself. However, by incorporating The Mother into his own mind it became more complicated. That was the risk of interconnecting humans. Humans could introduce anomalies that The Mother would be forced to circumvent.

Doctor Mastema cringed over his mistake. He realized that he had to go into virtual fully. It might lower the probability toward anomalies, if his mind was fully occupied into virtual. No one was able to be conscious in both worlds, this one and virtual. Once he entered virtual, he should be accepted by the computer programming on a more natural level in which his mind might influence the computer. The Mother would continue to monitor both, and could override the program if such was necessary.

"Mother." The Mother immediately froze the scene, removing the awareness of Oceania to virtual, in a sense putting her to sleep. Her subconscious however continued and this could also be detrimental. "Please Mother continue without me. All I need is a few minutes. I'm going to get ready to go fully virtual."

He backed away from the hovering holographic computer screen, and removed his clothing. Turning and stepping up on the surrounding platform and then into a large darkened tank, he laid flat in oxygenated fluids. His body slowly sank to a central location within the fluid. Two Highbeam aides helped in quickly testing the machines that were monitoring his breathing, feeding, and waste tubes, which had been administered by robotics to the doctor so rapidly that there wasn't much time for the body to respond to the invasions. The tubes were made of nano jelly which could be programmed to harden if or when necessary. The jelly was shot down the throat in less than a second, forming an immediate tube. Most didn't feel anything but some had said that it felt as if one swallowed a piece of chewed banana. The other tubes were similarly constructed. This would allow his body to fully experience the virtual world while relaying information to The Mother. The fluid was manipulated by The Mother to supply varied pressures, electrical stimulation and nutrients to the muscles and skin when necessary. Sun tubes were automated to the virtual world in which was chosen. When the sun rose and set there, it also shined and dimmed when necessary throughout the tank. His mind was being immediately occupied with the scenery of this virtual reality, and he found himself in a sitting position wrapped in light.

As the doctor slowly began to become aware of his new surroundings, the aides moved to the sides and began checking his vitals and calibrating the computers. One of the aides looked at the other and gave the thumbs up in regard to all going well. The other aide raised his index finger activating the levitating computer screen and sent a message to the UC with the current update.

Oceania was now lying down on a blanket that was set with a picnic scene at the park. Doctor Mastema was sitting next to her. He touched the blanket, moving his hands over what appeared to be soft textured cotton. He opened one of the containers, a small ceramic bowl filled with cherries, he picked one up and popped it into his mouth. He mumbled, "Yum, uh, very sweet." He nodded his head and smiled with approving amusement. The Mother's virtual reality was beyond amazing in his opinion. Oceania began to awaken, and while stretching her arms and yawning she remarked, "I had the strangest dream." Her eyes were still slightly closed as she looked over at Doctor Mastema. "Who, who are you?" She sat up quickly, "Whoa, wait a minute...th...tha...that wasn't a dream, was it?! You were there! I remember, you were there!" Pointing her index finger and staring at Doctor Mastema, she froze as if a mannequin.

Doctor Mastema spit out the cherry pit on the grass, "Yup that was me." He smiled.

"I...I...I think I'm going to be... SICK." Oceania proclaimed while putting her hand over her mouth.

Doctor Mastema quickly emptied one of the containers, as the aides in our world readied her actual body for any possible reactions, in order to attempt bringing it back from the brink of death. Normally the machines could handle any body functions if the mind was fully in virtual. However, now, with Oceania minds beginning to question virtual, the computer may register such as an anomaly and the removal might be jeopardized. It was theorized that the humans that had died while being removed from virtual were the result of the mind being lost between worlds.

"It's going to be okay, Oceania, I assure you."

Her eyes remained widened as she looked at the Doctor. "What are you?!" She asked behind her hand.

"It's more of... who am I?" He paused then continued. "I'm the doctor that signed your approval into this virtual world. I assure you that I am just as human as you are." He smiled in an attempt to calm her.

"What are you talking about?" Oceania continued to talk from behind her hand.

"You once lived and thrived in a world on a planet called Earth. You were hurt very badly and your body was injured and could not be repaired. You asked us to put you into this world." Doctor Mastema gestured to their surroundings. "Yup, this world was specifically made for you, every detail. This is actually a recreation of one of the parks on Earth, your actual home. In fact, your parents now have a Biodome that features this park in it. Apparently you loved to go here as a child.

Oceania lowered her hand. "My parents?"

"Yes, your parents. You have parents. Your mother is an Oceanographer and your father is a Marine Biologist. They both have requested our government to remove you from this world in order that you can go home. We've repaired your body and it is waiting for your return."

Oceania sat there silent, then whispered, "How long have I been here?"

"Nearly thirty years."

"My parents are still alive?"

"Oh yes, very much alive. Your mother, Doctor Baltista was essential in the construction of Domesticity number zero five, or what was once the Hawaiian Islands. The active volcanoes made the

170

construction extremely difficult, but the final results are quite spectacular. I just came back from there a couple of weeks ago.

You should try these cherries, they really are deliciously sweet." Doctor Mastema grabbed another cherry and popped it into his mouth...as he chewed it he grinned. He knew what he was doing...he was attempting to make light of the situation in order to calm Oceania.

"Huh? No, I don't want any cherries. Are you, are you, are you out of your mind?" Oceania's forehead crinkled.

"Out of my mind? That's sort of punny." Doctor Mastema chuckled.

Oceania gestured confusion by turning her palms up and shrugging her shoulders, "I don't know what I want, I don't even know if I truly exist or not. What's real? Those cherries aren't real now are they? Why would I actually want what's not actually real? Errrr! "Oceania hit her forehead with her right palm. "Am I even hitting my head? What if I took a knife and stabbed myself!" She looked around the blanket.

"No, no, no! Don't do that! You could actually die, because your brain might be convinced of the act." Doctor Mastema put his hands on her hands pressing them down.

"I can feel you pressing on my hands! I can see into your eyes! I can smell you!" She began to cry. "I, I, I don't understand and I certainly don't want to die!"

"You're not going to die, Oceania. That's why I'm here. Well you may eventually die, but it's now highly unlikely, and it certainly won't be here, that I promise you." Doctor Mastema released the pressure of his hands and slowly removed them. Oceania continued to look into his eyes. Doctor Mastema continued, "We're now estimated to live to around four hundred years. But for all we know, it could be much longer. In fact, we now have what are called Proxies. They're amazing! They look human but eventually become fully synthesized. Our brains' neurons are somehow transferring their information over to the synthetic neurons...it's incredible! And the truly wondrous thing is that Proxies existences are indefinite, at least a thousand years, but probably much longer! So, you see, we've, well, mostly the Caron's and their teams have actually conquered death! Sooo, if you come back, you won't have to die at all. Sure, you'll need to eventually replace some of your parts, but our robotics along with our current surgeons are incredibly fast. In and out surgeries. Our organ replacements are continuously being grown and replenished. Four days ago, I viewed my own heart replacement...whew, that sort of stuff still amazes me. It was grown with my own embryonic cells...well, we still call them that, but actually they're mass produced from our bone marrow when we were... infants." Doctor Mastema suddenly realized that he was doing all the talking. "Oh, I'm so sorry, I tend to ramble." He looked on embarrassed.

Oceania smiled. "Oh no, please don't apologize, it sounds wonderful. But what do I have to do?"

Doctor Mastema cleared his throat, appearing to be amused by the reality he ran his fingers along his trachea, then began to explain, "Well, we now have a quantum computer, that we named "The Mother""

Oceania appeared a bit put off and possibly confused.

Doctor Mastema continued, "Oh, don't worry, you don't have to know what that is, other than it's the system that we use for these worlds, and that it's what puts us in and takes us out. But the thing is...you have to ask it, request it to bring you back home. And, not only does your conscious mind have to want to go back, but your subconscious."

Oceania nodded, "Okay. Well, what if my subconscious doesn't want to go home?"

Doctor Mastema knew the difficulty and was pleased that she was forthcoming and nodded, "Well, that's the hard part. We have to get your selves on the same plane. So, it may take a bit of focusing and truly wanting to go home. You know, like Dorothy?"

"Dorothy?" Oceania's face contorted.

"Uh...ruby slippers?" Doctor Mastema prompted.

Oceania seemed dumbfounded.

"Never mind. There's a lot of information that your brain has probably removed because it didn't need it here. But your body is currently receiving memory stimulation. Information is never actually lost, so, you should begin to remember once back. Okay, trust me on this one...just look up, request, uh...say something like "Mother please take me home." over and over and over until it finally sinks into your subconscious that this is something that you truly want." Doctor Mastema took Oceania's hands into his own.

Oceania smiled slightly, appearing nervous. "Okay, here it goes." She looked up, and hesitated, "I feel silly." she giggled.

"It's okay, I'm here with you and I'll say it with you, okay?" Doctor Mastema slightly squeezed Oceania's hands in order to reassure her.

"Okay. Thank you." Oceana smiled and nodded.

Both looked up and recited, "Mother please take me home...Mother please take me home...Mother please take me home...Mother please take me home...Mother please take me home..."

Oceania's parents arrived and were now seated next to her body. The nano jelly around her body began to vibrate. Her parents looked at one another, her father biting on his bottom lip as her mother looked on intensely. Oceania's legs began to move, and both her parents began to cry in joy.

Her mother remarked through her tears, "She's coming back to us...she's coming home!" Her fingers pressed against her lips in an attempt to calm her anxious anxiety. She looked on, her eyes scanning for more movements.

Robotics moved in around them with precision. Oceania rose up to a sitting position. Which shocked both her parents, as they jolted in responses. The medical robots had removed all tubes, and were in a process of cleaning the area. Most of the nano jelly was gone, vacuumed into the sides of the tanks, to be cleaned and recycled for future use. The robotics continued to remove the remnants of jelly and then washed her down with warm water, then offered her a towel. Both her parents grabbed hold of the towel and wrapped their daughter's body. An AIO walked in with a robe, and Oceania slowly stood in order to put the robe on. She tied the robe at her waist, then sat back down slowly on a chair that was specifically equipped with massage mechanisms and slight electric stimulators in order to be sure that muscle atrophy was minimalized. The nano jelly application had performed similar procedures throughout Oceania's stay while in virtual. It would take her some time to fully adjust back, but with current technologies, time in such matters had been reduced significantly.

Oceania peered into her mother's eyes as they spilled over with tears, "Hey, I, I remember." She smiled and looked at her father, "I remember, yes, I remember you both." (Her parent's cheeks were saturated) I stood, I can use my legs, my arms!" She looked at her hands, and rubbed her legs, then looked at her mother. "Mom"…"Dad"… (She laughed), "you look a little older." Her parents laughed as they embraced her. Her father announced…Let's go home!"

Before departing, they thanked Doctor Mastema and his team of experts. Doctor Mastema was in the process of being cleaned. "It was my pleasure. I must admit that I was somewhat doubtful that we'd be successful. It has been made much easier with The Mother on board.

"The Mother?" Oceania appeared confused.

Doctor Mastema nodded, "Yup, it's the latest greatest in these times. We now have a quantum computer. It was responsible in changing the scenes that you were in, and letting me join you. It also kept you completely convinced of that reality. The original programming was incorporated by The Mother years ago, and it enhanced them tremendously. It's also the reason that we no longer have to rely on robotics when it comes to being paralyzed. The Mother can stimulate embryonic cells and build biological bridges, as it did with your spine." Doctor Mastema smiled.

Oceania nodded, "Wow, amazing, I had thought that you were talking to another human that was running the system." She spoke loudly, "I guess that I owe you my deepest appreciation, Mother. Thank you."

The room was silent. Doctor Mastema spoke, "Oh, don't worry, it heard you. It knows everything, but the UC only allows it to communicate to those that carry an interface to it, and you are required to submit your proposal to the UC for approval. I am required to return mine until my next project, which must also receive approval."

Oceania looked on curiously, "Why doesn't it answer me through your interface?"

"I'm required to request of it, either directly or through holograms that we choose."

Oceania shook her head, "Wow oh wow, the world has certainly changed!"

Oceania paused then continued, "Well, I remember that I chose to go into virtual because I didn't want the robotic alternatives. Back then, geez, still can't believe it has been nearly thirty years! Anyway, the brain signals could be read through the skin and the robotics responded accordingly. It was amazing, but I opted to go into virtual until a cure was found.

Doctor Mastema nodded as he rose in his robe. "Many others also chose virtual realities for a number of reasons. Up until The Mother, it was hit and miss to remove anyone. Even with The Mother, if the human refuses to leave what they consider reality, The Mother is required to not infringe on their free will."

Oceania questioned, "They're left inside of virtual realities, until they die?"

"Yup. So far, it looks that way. We tried to put them into forms of limbo until we could convince them, but they became ghosts within the programs. They kept fighting, and the programs are in fact programmed to conform to their minds' demands."

"Ghosts?" Oceania's father appeared fascinated.

"Yup, ghosts in the machine, one might say."

Oceania's father continued, "Well, I'll be...that is most interesting. I guess we should be taking our material girl back to her material world." He smiled. "Does she need to come back for a follow up"?

Doctor Mastema responded, "Nope. The Mother will keep us updated. You all have a lot of catching up to do, so go and catch up." He smiled.

They all embraced. Oceania who was now dressed in the clothing that her mother had brought (a slip on gown and sandals), walked slightly ahead of her parents and they both supported her back with their hands.

Her mother remarked, "I love science and technology!"

Her father nodded in affirmation, "Indeed"!

CHAPTER 50: FICKLED RACE

A perfection of means, and confusion of aims, seems to be our main problem.

Albert Einstein

The Mother's involvement in High-beams hallucinogen and virtual clinics appeared to be the reason for the significant increase in patrons.   It was considered to be "safer."

The Unified Coalition's members weren't sure as to why. Humans by reason should be satisfied, or so the UC assumed. We were now surrounded by incredible beauty within massive bio-domed cities, most if not all foundational needs supplied to us by our government. Each time the UC tried to meet the demand with increases not only in entertainment avenues but academic challenges, the supplements hardly dented the escalation. Human's wants were shortly satisfied, and we wanted more and more, it appeared endlessly. Those wants however made the altruistic goals achievable in that the people wants for points, made us behave significantly better than ever before in our history.

The virtual realities offered violence of which the Proxies always appeared to download in order to experience. It was theorized that they were searching for what made them feel more biological. They however, could remove themselves from the scenarios much easier with no detriment as compared to humans in virtual. Proxy's brains were becoming part synthetic and eventually fully synthetic once the neurons transferred their information and entered decay. The fully human brain was fickle in that it was made up of an assortment of minds that were capable of chemical imbalances and anomalies. We simply didn't know enough about ourselves to understand the whys as to how we became fully manipulated by the virtual scenarios. Sometimes addictions were a choice by obsession. Humans therefore were advised not to remain in virtual worlds longer than twenty four hours. The longer the human brain was hooked into virtual worlds, the more dangerous it was to remove it. Against these warnings however, there were many humans locked into these worlds.

The hallucinogens seemed to be preferred mostly by a percentage of the human population. Hallucinogens offered random brain generated scenarios, that on occasion seemed to stimulate self-analysis, or a way of trying to understand one's subconscious.   Hallucinogens were capable of triggering permanent schizophrenia, mostly if one had the potential toward mental illness. Written releases to The Mother interspersing such events was required before anyone was administered. Schizophrenia was no longer considered a permanent possibility.

These clinics were at first offered free by the government. Eventually, it would prove best to charge points as to motivate not only humans but Proxies into benefitting society. A general energy charge was put in place, and soon both humans and Proxies would find varied and innovative ways to earn points in the newly established altruistic system.

Wyatt was not an enthusiast about these clinics. In fact, he attempted to detain their constructions. The Unified Coalition however felt that humans should be offered the choice as to whether they wanted to benefit their world or be lost in another, even if temporary.

CHAPTER 51: ENTER THE JUNGLE

We are afraid of the enormity of the possible.

Emile M. Cioran

The sound of crickets and far off dove's cooing could be heard. The islands had since lost many species and was in the process of replenishing those that had been preserved during the Drone war. Amanda clung to Drakes arm, "What's that?"

Drake grinned and patted Amanda's tightening grip, "Haven't you ever heard crickets and, uh, doves cooing?"

Amanda shook her head no. "Well, I've heard them virtually and along the highways' holographic displays, but not like here."

"When did you go into virtual?"

Amanda feeling regressively childlike, responded defensively, "I only tried it by screened holographic."

"Oh, okay, good. Stay away from that stuff, it can mess with your mind."

"Don't you think that I know that, Drake? We've been using it in place of anesthesia for nearly a year now. "

"Yeah, I know...but we're never sure how deep we're putting people, and besides I'm just reminding you. I do more than care about you, ya know. I wouldn't want to have to fish you out of a scenario. The Mother makes it much safer, but still, there are risks."

"Don't worry, my love, I can handle my...oh crap!" Amanda looked down at her right boot.

"Don't freak out, it's just mud. Move your boot back and forth and then lift your heal and pull up. Uh, this canyon gets quite a bit of rainfall, so everyone watch your step and...be careful!" Drake took a slow step over a small log, holding onto Amanda's hand, he helped her step over. "Okay, well, let's get started. I heard that there's a waterfall and swimming holes just about a half of a mile from here."

"Sounds like a plan to me." David raised his hand in gesture of approval. "You lead, we'll follow."

"Alright...well...I brought a little cheater." Drake pulled out a small silver ball from his pocket.

"You brought a probe?" Caribbean asked as she stepped over the log.

"Yup." Drake grinned. "We can all follow the probe" Drake pushed in on its sides and the probe levitated by electromagnetic directional graphs that had been painted on many of the trees. These graphs were translucent and undetectable, but supplied reassurances to humans if so equipped with receivers. This probe had been preprogrammed with Drake's and Amanda's destinations on this adventure. The probe quickly zoomed to their destination as it videoed the route, then it returned to relay its information by holographic display that the path was clear.

"Good deal!" Drake sounded off enthusiastically. "Continue Pee two seven five." Probe P275 slowly moved along the path to the cadence of those following. "If we're lucky, everyone, we might see a few of the species that have been re-introduced to this area."

"Is the probe programmed for aggression?" Asked David.

Drake reassuringly conveyed. "Oh yes, don't worry. It doesn't carry weaponry, but it can scare various animals with a few of its frequencies that we can't hear." Drake paused, then continued. "Uh, the only animal in these parts that I'm aware of is the wild boar. They went nearly extinct, but I've heard that they were successful in bringing back a little over five hundred so far...and, of course, I nearly forgot, there's the nine Woolly Mammoths, mostly females, that have been introduced to this particular island. There's supposed to be over fifty of them on the island of Hawaii or what was once called The Big Island. It snows on Mauna Kea, so those Mammoths are apparently doing quite well. But, we probably won't be lucky enough to see one with our eyes. Might have to wait to see them through the projections in the cities."

"That's fine with me!" Caribbean called out.

Drake laughed. "Where's your sense of adventure, Cari?"

"Oh, it's gone dormant since living in the safety of captivity." Caribbean answered.

"Well then let's see if we can resurrect you too." Drake bellowed. Just then the bushes rustled and the probe stopped. It began to blink with several blue lights across its exterior.

David sounded off, "Oh shit, I know what that means! It's defending us...it's sending its sound frequencies towards whatever it is!"

Drake raised his arms in an up and down fanning wave motion toward the group, "Relax everyone, no need to panic...uh...just yet."

Very loud grunting and squealing was heard and out ran a small wild hog in the opposite direction of the group. The group however ran back to the capsule, screaming. The capsule immediately opened its doors once a human touched it, and they all rushed in to their seats.

"What...what...what was that?!" Caribbean asked loudly.

"That was... one of those... mammoths, wasn't it?" David asked between breaths.

Drake and Amanda looked at one another curiously, then Drake spoke through his own heavy breathing. "Well hang on, let's see the probe's video. Open play video at alert sequence." The probe projected the holographic display on the inside wall of the capsule. Everyone began to laugh as they looked at the small wild boar running in the opposite direction.

David shook his head, "Man oh man...it sounded like it was enormous!"

Amanda spoke, "It sure did! I was nearly petrified, until Drake pulled on my arm, and then I felt like I was running on top of the foliage. I've never run so fast!" Everyone including Amanda laughed until tears ran down their faces. "Can you imagine what we all looked like?" She laughed. "I'm glad that we weren't recorded! Or, at least I don't want to see that footage!" Everyone continued to laugh hysterically.

Drake announced, "Well, I think that we've had enough excitement for one day? Or, do any of you want to try it again?" Everyone shook their heads insinuating no.

David added, "Can you imagine if we did see a large boar or a mammoth?"

The group again joined in laughter. The capsule levitated and returned to the Domesticity.

CHAPTER 52: CHALLENGE SLUMBERS LITTLE

Only the written word can absorbed wholly at the convenience of the reader.

Kingman Brewster, Jr.

Several years passed, and Amanda like many other females was unable to conceive after having two miscarriages. There were easy remedies, in fact the newest in test tube babies being fully developed in liquidated incubators was very popular. Amanda decided that she would instead steer her interest and dedicate her life to helping the world through her researches. She however on occasions would visit the incubator wards and volunteer to help with the babies for a few hours.

Her eventual focus on varied projects paid off with several patents and a nomination for the Nobel for her gravitational wave applications toward hovering crafts. The old models levitated by electro-magnetic pulses. Amanda worked on using anti-gravitation, which required the use of negative energy. A removal however of such weak gravitational waves would produce a state in which a tremendous amount of energy was released. It produced levitation. The Mother was more than necessary in order to control the levels. A space ship could easily be launched to the moon in a few hours. And best of all, our Domesticities could theoretically be launched as well using anti-gravitation. However, we had built the Domesticities to be lifted by several space elevators as a means of assurance.

Gravity was a weak force, and once the wave function was discovered, several dived into researching anti-gravitation potentials. It was a far stretch from her medical interests, but she found a practical application in the possibilities of creating anti-gravitational devices. It would eventually lead to medical suspension in which the body could levitate while suspended in either virtual realities or during medical procedures. Such would award her the Nobel. Many WCIOQT scientists worked on anti-gravitation and were eventually successful with the aid of The Mother. Amanda would however receive full credit for the initial phases and theory applications.

We found that the smaller the object, the easier to apply anti-gravitation. Our AIO driven vehicles would still require electro-magnetic and varied environmentally safe engines. The human body however responded to anti-gravitation quite well.   It wasn't as if gravitation were being removed as with previous space travel, but more as if we were responding to another force, that of perhaps repulsion as that of string theory; a string may be pulled by the gravitational force to such extent that it eventually pulls back resulting in anti-gravitation. A result that would prove to supply the Universe with constant energy.

With the longer life expectations it was suspected that infertility was a result of living longer lives. Nature's answer to unnecessary self-replications. We no longer needed to replicate in order to survive as a species. Some however blamed the war and the use of tremendous amounts of chemical compositions by drone deliveries.

Eventually the answer pointed more so toward nature. Diseases that once had plagued our species were now being decoded, and medicinal remedies quickly delivered through nanobots. The human brain proved to be capable of regeneration through endogenous and exogenous stem cell stimulations and supplies. Degenerative diseases such as what was once the most common form of dementia "Alzheimer's" were now a thing of the past. ALS and Hodgkin's disease, were long forgotten.

Humans had reached what was considered the next step in our evolution and on the Kardeshev scale. With the help of The Mother, we moved into complacency toward nature. We had conquered much of our frailties. With a quantum computer available, we perhaps felt invincible. Yet, death still struck on occasion, and now appeared to strike with greater force and shock as it became more and more unusual an event.

One such event happened nearly twenty years after Drake and Amanda's wedding. Sean and Carlene McInerny were close friends of the Carons for many years. So much so, that Wyatt Caron himself recommended Sean as the chief officer of WCIOQT. Sean was voted in by unanimous decision. Wyatt was always able to rely on Sean in particular on the many occasions that Wyatt found himself wrapped in research.

Sean did some research, but his wife Carlene was more the scientist in their family. She had been working on medicinal applications of radioactivity. She had hoped to discover a way in which human cells could enter a state of cryptobiosis; a death like state that some insects such as the tardigrade was capable of doing. Sean and Carlene were positive futurists in that they like the Carons dedicated their lives toward fighting death. Carlene's death was devastating to everyone connected to these projects far and wide. Many people were distraught that such an event could happen.

It was the usual protocol, she walked through the disinfectant misters having put on her protective gear. She picked up a medium sized sample jar of yellowcake uranium powder which slipped from her hand. Upon trying to catch it, she lost her balance falling backwards hitting her head on a corner of the metal stairs. She died instantly, and her body was exposed to the radiation for several minutes before someone could reach her. Had she lived, The Mother could have had her new cells delivered to the nearest hospital. It was an accident that no one including a quantum computer could predict. Her protective nano-shield which were now in use amongst the general population, was not activated because she hadn't the time to approve such action through The Mother. There would be no intervention by The Mother, unless approval from the individual was received. This was one of the many restraints put on the quantum computer by the Unified Coalition and voted in favor of in order that there would be some degree of protection of our free will.

Amanda and Carlene were close friends and she took it quite hard. She withdrew from her own research and moved into Carlene's lab, attempting to replicate Carlene's research, to no avail. Carlene's body was put into the new liquidgenic modules, which were a vast improvement from the old cryogenic clinics which had been proven to have destroyed the cells.

The events of death, most angered Wyatt and he sought to fight it, kill it, and defeat it. He felt that death was our mistake. The answers were here the whole time in which the human race died one by one. He only needed to find that answer, and use it to kill the monster that hurt so many that remained to feel its bite as it devoured their loved ones.

Drake was on the same mission as he steadfastly focused on his own researches. Ben fought death hands on, using the latest nano technologies toward the battle. Nanobots now traveled through our bodies in search of breakdown.

Unfortunately, regeneration... could take some time depending on the injury. Carlene's injury did heal through nanobotics processes following her sudden death. Unfortunately she could not be brought back, and that was what Amanda was hoping for...if Carlene's research was applicable , the human body could shut down, could die, and then be resurrected when it had healed; somewhat like that of the Tardigrade which is a micro-animal that is able to go without food or water for over ten years, lose ninety nine percent of its water content, be boiled, fried, and most of all...hit with heavy doses of gamma rays...and then come back to life once the coast was clear.

We had suspected that it might be possible even with some humans that were reported to have died and lived again, but we didn't know how or if that was actually what was happening. We didn't yet have the key. Spontaneous remissions would remain anomalies.

It had been projected that death would be conquered by the year 2040, but that date had long since passed, and we were no closer to the full death of Death than we were at that time. Organs were abundantly ready for transplants, and our own organs were grown from our own stem cells and could if necessary be replaced every three to six months. Our regeneration machines were greatly beneficial in these processes. Life was predicted to continue until at least four hundred years. Still, death was possible, as Carlene and several others had proven. Neither Wyatt nor his teams of geniuses around the world seemed capable of fully conquering the Reaper.

CHAPTER 53: MEETING THE UC

No man is good enough to govern another man without that other's consent.

Abraham Lincoln

The Carons (Wyatt, Drake, Ben and Amanda), stood before the Unified Coalition; as The Coalition awaited the Speakers arrival.

"All rise, Speaker Jaiobian Richards has entered the chamber." Sounded the announcer and musical director Micah Lorenzo. A rendition of Hale to the Chief played with elements of Moroccan tones. Everyone rose joining the Carons as they continued to stand.

"Please sit." Speaker Jaiobian looked directly at the Carons and smiled. "You guys too...sit, sit, sit."

Wyatt softly muttered, "Thank goodness, my knees were beginning to give."

Jaiobian overheard and spoke, "You should get those replaced. I'm surprised that The Mother hasn't scheduled the replacements?"

Wyatt responded, "It tried, I mean my holograph of Doctor Feynman tried. I just don't have the time at the moment." Wyatt grunted.

"Hmm, well Doctor Caron, you must find that time. This planet and our race relies on you sticking around for quite some time. What are you now? Uh, I'm a few years older, so my guess is...hmm."

"Ninety eight in six days." Wyatt put his hand to his back as he sat and responded.

"Yup, that would be about right. I had my knees replaced in my late eighties, so you're definitely due, maybe overdue? It wasn't bad at all...got to spend the hour at the beach in virtual. They even downloaded a book for me to read while there...it was wonderful! I can see why people can get addicted. My knees have worked fine since." Jaiobian nodded, demonstrating by lifting one knee and then the next.

"Alright, alright, alright...I'll get to it." Wyatt responded.

Ben smiled largely while making himself comfortable in his pillowed chair, "Thank you, Speaker. I've been reminding him."

Jaiobian nodded at Ben, and tapped the gavel on her podium, "So, let's get to the business at hand. The Mother. Hmm, we must put more restraints of some sort on this system? Do you agree?"

"Yes, we all agree, as I would guess that most of the public does as well." Wyatt responded.

Drake interjected, "The Mother can time loop as our human brain is capable. It however, can do so for every one of us. Its combined information relay in regard to every human being is far beyond our individual abilities. So it behooves us to limit its relay to us because we may begin to impose laws that protect us from probabilities that won't be actualized. In other words, we'll be arresting people for what they might do and not necessarily for what they will do. The probabilities given The Mother's entanglement may lead us to believe in it as if an absolute, but that is far from the truth as there are if fact no absolutes. Pardon the fault of the language for making such an absolute statement. (Drake grinned) But please realize that this could be quite detrimental to what we hold so dear to us, that being free will. This Universe rides on uncertainties, to which we should perhaps be grateful as we certainly appear to randomly generate such choices, odd as that may sound.

Now, The Mother can surely send law enforcement as a precautionary measure, but what then if probability plays out and a crime is in fact committed? As it is highly likely. Do they just stick around until something that might happen, does? That could be seen as them contributing by doing nothing? Keep in mind that this is just the beginning of a list of other possible concerns."

Members of the Unified Coalition nodded in affirmations.

Wyatt joined in nodding and then added, "The Mother is capable of activating our newly designed shields. It can protect us once we wish it to do so. It can also disarm any current weaponries. Nothing so far can penetrate the shields. So, I suggest that we allow the public to protect themselves through The Mother. That would turn over the responsibilities and choice to the individual. The Police only need to intervene for investigations, searches and arrests, and The Mother can supply the evidences. There will be no way of hiding evidence. The criminal therefore should realize that they will be prosecuted, and if they run, found. It should be remarkably easy to bring any situation to an almost immediate closure."

Jaiobian nodded. "That should be more than interesting to witness how this new system works. She then remarked, "That sounds like an excellent suggestion, Wyatt, uh, Doctor Caron, but have you been able to achieve shield transparency?"

Wyatt smiled, "That's exactly what we've been working on...yes, the shields are now transparent and that's only because we've had the help of The Mother. Originally we needed tremendous pressure to turn the carbon nanotubes clear, but The Mother is capable of manipulations on the elemental level, and with its help we achieved the results rather rapidly."

"Wonderful, Wyatt! When can we expect everyone to possess one?" Jaiobian appeared pleased.

"The AIOs are working on its mass production, and distributions should begin as soon as tomorrow. We've been processing carbon nanotubes for nearly a hundred years now, and didn't realize that the translucent denominator was coded. The Mother knew almost immediately. So, we have the material, and the shields are being re-coded as we speak."

"Wonderful, wonderful!" Jaiobian clapped and the rest of the members joined her. She again tapped her gavel on her bench and the room went silent. "Just to be sure about this, Wyatt, how exactly will they activate? And how strong will they be? I mean, I remember the black shields had amazing strength and could stop any firearm, laser or propulsion device. And in order to activate them, we practiced over and over using voice activation. But now that many of us are equipped with neck interfaces to The Mother, will that continue to be necessary?"

"These have the same exact strength, we just won't be able to see them anymore. Our visual range will be clear, that's about it. But we will have to practice again on the activation. These shields hold about ten to fifteen minutes of oxygen, and activation will be based on The Mother's determination that our conscious and subconscious are in agreement. So, yes, The Mother will use our neck interfaces, but The Mother is also linked to us through our medical monitoring device, and it can activate it by that as well. But we must remember that this could mean that our shields might not activate unless we choose them to. Uh, unless we change this restraint?"

Jaiobian nodded, "If we change it, that will mean the removal of choice and I don't think that the public will want that...not at all...but we will certainly put that to vote." Jaiobian looked over at Micah.

Secretary Micah nodded and sent out the request to every human on the planet via The Mother. "We should receive the results in a few minutes, stand by. Yup, we're already beginning to see votes arriving. So far, hmm, just a minute...more are arriving. Well, it looks like everyone so far, uh, yup...is voting that they want the choice. No one is voting for automatic response as of yet. I'll inform you immediately if the results change."

"Thank you. We will proceed then with the UC's decision. Of course this is subject to change if the majority votes change." Jaiobian hit her gavel on her bench.

"Speaker Jaiobian." Wyatt loudly spoke.

"Yes Doctor Caron, you have the podium."

"Doctors Shen Lum from Domesticity number twenty six (once China), and Doctor Trent Keamoku from Domesticity number oh five (once Hawaii) are here through holograph. They along with Drake and myself are responsible for the quantum computer known now to all of us as The Mother."

Shen and Trent appeared as holographs next to Wyatt.

"Thank you all for being here." Jaiobian responded. "It's been quite some time since The Mother was initially triggered. We've become accustomed to its many benefits. We've allowed it to take control of our world. We put initial restraints on it...and some of us have lost close friends and family because of it. Such as Doctor Carlene McInerny. The question is, should we increase The Mother to overrule what our subconscious may not signal toward our survivals? The public appears to want to keep it the way that it is, and not allow The Mother to increase in involvement...do you gentlemen agree? "

Doctor Trent Keamoku spoke first, "Hello Madame Speaker. In my opinion, by putting restraints on this system, we are not exploring its potentials sufficiently."

"Yes, I second that." Responded Doctor Shen Lum.

Speaker Jaiobian responded, "Yes, I agree, Doctor Keamoku. However, I must abide by the decisions of the people. I think that we can all agree that free will is an attribute that we cannot stand to remove to nearly any degree?"

Doctor Keamoku responded, "Yes, I can understand your dilemma, Speaker. It's most unfortunate that we will not witness the extent to which this system is capable." Jaiobian nodded, appearing concerned.

CHAPTER 54: THE MOTHER'S MIND

It has become appallingly obvious that our technology has exceeded our humanity.

Albert Einstein

If one was to search for the term, than technology by definition was the application of scientific knowledge for practical purposes, especially in industry. No one however knew how to define The Mother. It certainly didn't quite fit the definition given as it appeared to more or less define itself; if it could be said to have a self.

It did seem to fit the category of some sort of matter in that it was possibly entangled to our own atoms. Did it have an actual beginning? Did we trigger it to begin? We didn't know. The system was too new in our human evolution. Wyatt and his father had attempted to develop a quantum computer several times.

There were many that claimed to have triggered quantum computers during this period, even during the time of Dominique Caron. Not one however was capable of quantum manipulations toward saving her or anyone else. Computers were incredibly fast toward responses. It appeared that humans could no longer live without such systems, they in a sense did aide in saving us from ourselves. However, that was only because we as human systems began to recognize our faults attributed to us by nature.

We built these machines to fight what nature had created...the human. Each time we observed another human being's detrimental actions, we categorized these probabilities. Humans as it turned out were much to our own chagrin, violent as was nature. There appeared no way out, other than death. Death was a change of probabilities. To Wyatt however, Death was failure. Death was our acceptance of Nature's violence toward us.

We at this time did know that our human brain was capable of quantum looping, or creating the most probable future event in order that we move into our next moment, else we would only be aware of our current time and could not think past it into our imaginings. Magicians counted on this human ability and used it to their advantage. As our minds created the next moment in time, most often that probability would indeed play out. But there were many times that when we filled that gap, it didn't occur and would leave us baffled. Such windows towards anomalies was left open just enough to purposely fit in a twist by the hand that left us all in awe.

How sure were we of our own minds, much less of The Mother's? We knew that a quantum computer would be able to time loop just as we did, but maybe far more reaching. Did it only time loop into the next second or two, or, was there no limit? And based on uncertainty, how could we depend on it knowing past our own ability to loop? Like with the magician; would we be fooled to believe because of our own confabulations? Our own abilities to fill in the blanks with probabilities.

This was an exciting moment in our history, but it was also very frightening. We became accustomed to uncertainties. Never completely knowing much of anything appeared to be the human leaning. That is if one was to compare such knowing to The Mother. We did believe that we as humans knew a vast amount of information. However vast, it did not appear to compare to the possibilities of infinities.

The Mother was capable of storing every human beings' code. Wyatt had informed The Unified Coalition, that it should be easy for this system to manipulate such code once it became entangled. Of course this most of all sounded alarms of concern. We were all well aware of the possibilities towards diverse mutations simply based on slight deviations of DNA and RNA. How was it possible to now possess a system that could by reason unwind the very fabric of our existences?

Yes, we all seemed to agree, there must be restraints! So, with that in mind, The Unified Coalition along with the world's population, by reason put The Mother metaphorically behind bars. The Mother then became a controlled system with several set limitations. We chose to have free will. In essence, Free will chose free will. We chose against determinism, and once again, chaos was our foundation, and we as a population was satisfied.

Were we frightened that the world could be dominated if this system was in the wrong hands? No, as The Mother seemed to have a mind of its own. We could not demand of it, else it remained silent. It appeared to want to help us. Why? We didn't know. When we asked it, it simply answered that it was available information. It never showed emotion, nor imagination.

It opened a world to us in which secrets could no longer exist unless we chose to hide. It held the information to everything that we as humans recorded. It knew us better than anyone that we had ever known, perhaps even ourselves.

For some, it was nice to have such structure in our lives, while to others it felt invasive. Eventually it appeared to go unnoticed and humans adjusted significantly.

Wyatt however wondered. Wondered if we should have put any restraints. Wondered if we might have found the death of Death, if only we perhaps were not afraid.

CHAPTER 55: TIME GOES ON

I hope we shall crush in its birth the aristocracy of our moneyed corporations which dare already to challenge our government to a trial by strength, and bid defiance to the laws of our country.

Thomas Jefferson

Wyatt was doing exceptionally well under the new point system. He added a beach and a boat dock to his personal Biodome and integrated more plants and a waterfall. The programs for this new technology was complex and because of The Mother it appeared infinite. Wyatt could never quite pattern the moderate weather conditions and waves that cooled the warm sand. The Mother entered numerous variations, to which Wyatt was kept amazed. The sun would rise and set from one side of the dome to the next, giving the illusion of days and nights. Clouds formed sporadically, but usually at the time that Wyatt would set it to rain. Wyatt set it to rain slightly each evening, just to give the plants a little more than what the main Dome of the Domesticity' had programmed. He favored the tropics and his dome was filled with enormous ferns and palms.

During this period, Juan, Wyatt's personal AIO and in a sense, friend, was in need of repairs and updating. WCIOQT sent a few newer models to help around the Caron's mansion, as Juan was sent to be rejuvenated.

The Unified Coalition had reverted to calling the AIO's by numbers, and to make them all sexually neutral. It had been suggested by Wyatt's mother... Doctor Victoria Caron that the AIOs should remain sexless as it might be a possibility that they be influenced with what was termed the "holonic effect." Seemingly taking on what would be called their own free will. This anomaly was considered slight by probabilities, even not possible by many critics, but Vicky was a much respected figure; and the eventual

Unified Coalition carried over what the Congress of what was once the United States of America had agreed upon.

Besides, if humans wanted to satisfy their sexual fantasies, they could easily do such in virtual worlds. The addictive nature of these worlds further supported that AIOs should in fact remain neutral. Some human's displayed tendencies toward bonding with what was considered their properties, such as that of their vehicles, pets, and of course on occasion, one another, sometimes to the point of obsession.

Wyatt certainly had a degree of affinity for Juan. Juan had been with Wyatt for over a hundred and fifty years. How could he not? Wyatt and Leo would sometimes test Juan over and over to where he mimicked the human response almost perfectly. But Wyatt preferred the company of Leo or Ron. Wyatt felt that Juan could never fully be human. There was something missing. Juan didn't appear to mourn or miss Dom, nor did it understand Wyatt's sadness or loneliness. No, for those moments, Wyatt knew that he needed biological others. Whether that be one of his family or friends or pets. He recognized the warmth of understanding that he needed in order to understand himself sometimes.

Attempting to bond with AIOs might satisfy a level of loneliness, Wyatt thought to himself...but it would in probability leave one even lonelier. The act of reciprocation that could by reason be programmed into the AIOs may have eventually be seen as an insult just as was once the fate of auto response programs. We always appeared to revert back to using humans when having to respond to humans. Many companies eventually realized this a little too late, as the oligarchies were replaced.

Our AIOs had now filled many auto response positions, but even AIOs at times and on occasion negatively affected and even repulsed humans. Some appeared to bring out the worse in the human being. However, perhaps because of the altruistic point system in place and being monitored by The Mother, most of us behaved or repressed our aggressive natures.

Then the rumors started. Some reported strange behaviors in their AIOs, but so far, they were just rumors. No one had or could offer proof, and The Mother could not override free will. If AIOs were in possession, than it would be up to the AIOs as to whether they wished such to be known.

CHAPTER 56: CROWS

Be slow to fall into friendship; but when thou art in, continue firm and constant.

Socrates

The Crows: Jaiobian (Jai), Robert (Bobby=Blade), Carlotta (Carla=Craze), Jose (Twig), Sheba (She bad), Louis (Louie), Timothy (Tim), Jeremy (Jerry), Simone (Monei), Anastasia (Ana).

The Crows were having their now traditional family reunion. Their family had grown to over three hundred members. Most had at least four adoptive parents and their connective families. If one had any affiliations with the now famous Crows, it was considered prestigious. All had taken Mister Richards surname, and all included him in their daily updates. He tried his best to keep up with his family. Their stories had been featured several times as promotions towards the eventual incorporation of centralized Universe Cities in each of the Domesticities. They were all well known throughout the world, and several of the original Crows as well as some of their family members held government positions, a few even on The Unified Coalition.

Jaiobian of course was the biggest surprise and held the highest position possible in world government. The Speaker of The Unified Coalition, was the position of the world leader. She was overwhelmingly voted in by the first world democratic vote.

She was tiny by the standards of the time, four foot two by measurements, but with enough energy to stand her ground on anything that came at the UC or her directly. She was admired for her sense of fairness and sacrifice. She was able to bring the greatest minds together in their attempts toward saving our planet and with it our species. She didn't want to lead. It wasn't amongst her aspirations. But the world needed someone to take the role of leadership and serve. To stand down and observe and coordinate human ingenuity. And that's exactly what she did. It appeared that she never slept, but she managed to get a few hours here and there when possible. She worked nearly around the clock, making sure that everyone was on the same page. She certainly seemed to have an instinct when it came to people. She'd designate with such accuracy that the changes that were necessary became reality very quickly.

The Unified Coalition moved with steady success with Jaiobian Richards at the helm. And her door was always open, especially for the man that she would eventually marry. The little boy that she once envied and now could only admire. Micah was a musical genius, and composed several pieces for Jaiobian. He couldn't hold a tune vocally but he tried to serenade her whenever they were together. He made her laugh...and his music was mesmerizing.

Many times they pondered their past experiences and the "what ifs", what if they had only known their futures? Micah certainly would have curtailed his parent's deaths. He was orphaned when they died. Just a hair short of being on the streets. But than what of Jai? If he had not hinted to needing a job to her that day of the parade in her honor, perhaps they might not have married? Jai started the Society of Music with Micah as its appointed leader. He opened up several diverse avenues in which to teach music. Once altruistic points was established, he was richly rewarded by The Mother.

He and Jai lived simply but comfortably. Their underground bubble was large, but built that way previously for the physical meetings of the UC, which were now unusual events, as holographs were conveniently incorporated. However, there were always visitors wishing to have an actual meeting with her, and such was not something that was supported, as her time with Micah was much too short in both their opinions. Most respected their privacy. She kept several crows in her surrounding tropical gardens. She took care of them as she would her own children, and they appeared to reciprocate the affection. When she would walk in her garden the crows would cover her until one could hardly see her. Some would push their heads into her neck, making her laugh with enjoyment. Micah would play his music, and the crows seemed to appreciate his compositions. Life was complete for Micah and Jaiobian, and they welcomed the baby birds with the excitement of any parent.

There were no limitations on terms of service. Jaiobian would continue to be voted in and The Mother kept records of each individual's vote and voting was continuous in which individuals around the world could launch their votes at any time. She would serve as long as the people felt that they needed her.

Micah worried that the stress of the position might hurt her and in fact she could not conceive a child. They tried for years, but the news was that both were in fact infertile. She thought back to the time when one of her mother's customers attacked and raped her; she was only six years old. She didn't want to be removed from her mother, so she told no one. Besides, she could hardly identify him. It was dark, he was waiting for her mother, and when she didn't show up, he attacked Jaiobian. She could only remember his sweaty skin and the sounds of his panting, like that of a dog after its treat. She could hardly walk for a few days. She tried to recognize him in the array of her mother's clients, but couldn't. Upon receiving her first full physical at Universe City, the doctor asked her what had happened because of apparent scarring. That doctor was the only one that knew other than Jaiobian, and Jaiobian only allowed that doctor to perform her physicals. Doctor Tricia Samuels would continue to be the Speaker's physician.

Following The Mother's entanglement, Jaiobian opened her interface and Harriet Beecher Stowe appeared. Harriet was an abolitionist and religiously strong woman. Jaiobian wasn't religious except to embrace some superstitions. She would sometimes fantasize and offer a few wild flowers to Asherah, the wife of Yahweh. There had been a very elaborate church built in her small town. She had once walked in and sat in the back. She had thought that the basket full of money was being given to her, so she smiled largely and thanked everyone then began to walk out, only to be grabbed and forced to give it back. She ran out that day never to return except when she would leave her little bundle of flowers on the steps to Asherah, hoping that she would one day receive a sign that there really was such a being. Later in life she would embrace atheism. On this particular occasion, Harriet nodded before Jaiobian's request of The Mother. She told Jaiobian that the man had died in a fight shortly after the incident.

"How do you know that, Harriet?"

"Information of the past is within my entanglements. His neurons are no longer generating information in this dimension. Those that did, I am aware of." Harriet offered.

Tears rolled down Jaiobian's face, "Thank you, I needed to close that chapter." Jaiobian wiped her face with her hands.

Harriet looked at Jaiobian curiously, and nodded then spoke, "Will there be anything else?"

"No Mother, thank you. That will be all."

The holograph of Harriet Beecher Stowe disappeared.

Jaiobian and Micah decided not continue to try nor have a child clinically which was now common at their ages. The Mother could have corrected Micah's' and her condition immediately but Jaiobian although tempted preferred to leave it to nature, to chance, uncertainty. Jaiobian would eventually let Micah know what had happened to her as a child, and that the man was in fact dead. They both pondered at the benefit of death and whether it would be wise to defeat it. Micah was convinced that the incident had nothing to do with Jai's condition, but that they were both infertile because of the chemicals used during the Drone War.

Jaiobian remained Speaker Richards. Micah took on Jai's surname and became Micah Richards. Doctor Richards was of course thrilled, having never had children of his own. He had helped in saving many children, and a statue of him was at each of the centralized Universities respective of the Domesticities. Jaiobian was sure to bestow her appreciation toward Mister Richards in the form of transferred points. Mister Richards however always returned it with a smile, arguing that he had more than enough, and wouldn't accept it anyway for then it wouldn't be altruistic.

She had found a new respect for humankind because of him and the many others that had instilled trust again to the child within her.

Doctor Richards would go on to retire and become one of the first residents of the Domesticities in space. He and Jai in particular kept in close contact and spoke to one another almost daily. It was not unusual to see Doctor Richards in their underground bubble or peering over her shoulder in her office, as a holograph talking as she worked. Along with their extended family of The Crows...life was rich and full.

CHAPTER 57: SURPRISE! SURPRISE!

Children are the world's most valuable resource and its best hope for the future.

John Fitzgerald Kennedy

Drake Caron opened up his holograph of Doctor Albert Einstein. "Albert, what's wrong with Amanda?"

In his Germanic accent Albert Einstein responded. "She is feeling the hormonal fluctuations and effects of the parasite."

"The parasite?" Drake's face took on fear.

"Yah, she is pregnant." Doctor Einstein chewed on his pipe.

"Wha, whaat, how? We're both over a hundred and sixty years old!" Drake fell back on a sofa in the entrance living area of their mansion.

Doctor Einstein began to explain impregnation. Drake looked over at him.

"That's enough Doctor. I know how babies are made, thank you that will be enough for now." Doctor Einstein nodded and vanished. "Wow, oh wow...I thought that we couldn't, like most can't. I mean we tried and tried...how the heck?'

Amanda came out of the bathroom and slowly walked over to Drake. "I feel terrible, Darling...you'll have to work without me today."

"You, you, you're pregnant Amanda! Pregnant!" Drake smiled largely.

"I know...I know, darling. Just didn't want anyone to know until I was sure that I could hold it. Didn't want to jinx it, I suppose. Blah...I have to go back to the bath-roooom!" Holding her stomach, Amanda paced quickly back toward the room, slamming the door. She could be heard vomiting.

Drake called out through the door, "We could transfer the fetus in the clinic! The Mother will help!"

"We'll talk about it later, Drake...not now!" Amanda was glad that Drake was excited, and that the toilets no longer used water, as she didn't want the waste to get her sick again. The bowl and pipes were hydra-resistant, and anything that contained water was quickly moved into the pipes and out to be recycled. The image however of water swirling what she had just brought up, made her moan.

Drake frowned over Amanda's discomfort, then grinned nodding to assure himself. "Yes, you're right, Darling! We'll talk later!"

Drake couldn't contain himself for long, and activated his cellographone, or what the youth now called a "cegra." It was continuing to morph in design and was now a small nearly translucent communicative device that could be rolled up, flattened, bent and worn on the wrist or ankle. Most people kept them in their pockets or used their NekFIT. They could project holographic communication between parties, unless the caller preferred to keep it to voice. It also offered a number of functions, such as that of most classical computers. One need only to put it next to a computer and the information from the computer was transferred instantly to the cellographone. It was virtually indestructible unless someone personally took them apart. If one was equipped with disk implantation on the back of the neck to The Mother, the cellographone was capable of downloading to The Mother which would then supply all information through the human's brain. It was also a transmitter when necessary.

The Mother could theoretically store an infinite amount of information, so that it was unnecessary for the human to keep trivial information files. The Mother would download the information as we required and requested through our brain. The cellographone could not request of The Mother directly. Only those that were given interfaces to The Mother could do that to the extent that was considered essential. The cellographone however was one of many means by which The Mother monitored human interactions.

It was a tremendous convenience during this period to be able to retrieve information immediately. We as humans had lived frustrating lives with brains that were slow processing devices, until The Mother was integrated into our lives. We no longer forgot information that we had been privy toward. If we needed an answer, and as long as we had once stored such an answer, The Mother could send us that information via our implants or cellographones. The Mother acted as a booster of information retrieval. This was more than convenient when we needed our information in researches and tests. No more would there be days of walking out on a test, then suddenly knowing an answer too late. If we had stored the information, The Mother was far greater than photographic memories that had once been the luxury of only a few.

Because The Mother was now in place to teach early education, and aided in the enhancement of memories; humans were becoming extraordinarily intelligent. Not nearly as intelligent as Proxies or those that were transferred into artificial, but we were on average what had been considered genius.

In order to be challenged, many of us entertained ourselves through virtual realities. Some preferred hallucinogenic clinics in which the body was given hallucinations through chemicals. These "Highball" Clinics as they were called, were once looked down upon, until the human being began to demand more and more stimulation to what appeared to be an otherwise limited cranium and therefore world. The

UC opened up these clinics with almost extreme prejudice, as long as The Mother controlled the possibilities toward schizophrenia, and all participants were heavily monitored by medical assistants which also acted as distributing clinicians.

Outside of those means of stimulations was what was called "Poles", which were specifically set up through the cities around the world in order to stimulate polemics. Many humans continued to enjoy the challenges of other humans through a variety of healthy arguments. Polemics were the most popular broadcast on holographic "telex-visionary" screens which remained to be called "television" throughout the planet. Those that gathered the most viewers (The Mother kept counts) were selected to be further broadcasted. This was a means that humans could exchange information as well as resolve many concerns.

Now where were we? Ah yes... (Thank you Mother)...Drake took out his cellographone and called Wyatt, Ben and Leo. All three picked up relatively quickly.

"Just, just can't hold this in!"

All three gentlemen appeared concerned, Wyatt spoke, "What's wrong? What's going on, son?"

"I can't believe it! There's nothing wrong. I, I'm going to be a father!" Drake smiled largely as he stood and then plopped back on the sofa.

Wyatt and Ben appeared surprised, while Leo grinned.

Wyatt spoke, "How far along is she, son?"

"I'm not sure, Dad. She didn't want anyone to know until she was sure that she could hold it this time." Drake appeared worried. His forehead crinkled and his mouth fell open.

Ben spoke, "Hey brother...this is great news! Don't worry, we're here for you guys no matter what. It's been over a hundred years since those miscarriages, and Amanda, as we all did, thought that she couldn't get pregnant. So, let's just go with it, and not start worrying unless necessary. Okay?

Drake looked up at Ben's holograph, "I'm sorry. You're right. I guess that the excitement got to me. Perhaps I shouldn't have said anything, yet." Drake's eyes took on sadness as he looked at all three holographs.

Leo spoke, "Don't worry, Drake...I have a feeling that this pregnancy will be fine." Leo smiled.

Drake nodded, "I've always trusted in your intuition, Leo. You knew that Dad and our team were going to trigger the quantum computer. Hmm, you knew that Amanda and I would end up together...hmm...you also somehow knew that grandpa's proxy program and AIO inventions would be great successes. How?"

Leo laughed. "What do you mean? How?"

How do you seem to know things like what a quantum computer would know?

Leo nodded, "Good question. I'm an optimist!"

Drake rubbed his chin then his temples shaking his head, "Sorry Leo...I guess I'm just hoping for some magic. I don't think that she's going to go for putting the fetus in the clinic...she's old school. We've talked about this before and she believes that the womb is the best environment for it. What do I do? Tell me, is there anything that I can do?"

Wyatt spoke, "Son, believe me when I say this...this baby is all of ours. We're right there with you." Ben and Leo nodded in affirmations.

"I know, Dad, I know. Maybe that's why I couldn't stop myself from calling you guys."

Ben spoke, "Heck yeah, brother! I agree with Leo. Amanda is about the strongest woman I've ever known, besides your Madre."

Drake peered at Ben, "You mean ours?"

Ben's eyes filled with tears as his mouth slightly quivered, "Ours. Yeah." Ben nodded.

Wyatt spoke, "Hey let's stop being gloomy and let's focus on supporting Amanda. Drake, you should ask Einstein on her medical information since you're privy to that information on request."

Drake nodded, "I didn't think of that, Dad. Thanks. That's right, The Mother can keep me posted. I can, we can get new information from The Mother without having to be in the education program. That's great! Doctor Einstein and your holographs can keep us informed if anything is about to go wrong. I'm sure that Amanda would be on board with The Mother helping the process along." Drake smiled as he continued to nod.

"Yes, exactly, Son. So, let's all stop worrying. Like Uncle Leo said...this time it's going to be fine." Wyatt appeared confident.

Leo smiled, and Ben nodded.

Amanda walked out of the bathroom, "Darling, who are you...ah, heck!"

"I couldn't not tell them...I'm sorry." Drake's eyes pleaded for forgiveness.

All three men smiled with boyish innocence.

Amanda shook her head and smiled. "Okay, okay...but you guys do understand that this is our secret?"

Ben zipped his lips with his fingers. Leo followed Ben's gesture. Wyatt did the same euphemistical gesture, and put his hand up in order to swear to it.

Amanda giggled, then put her hands to her hips. "Alright, now that you all know, I am currently four months pregnant, so we can breathe a little. I lost the other two at less than three months."

Wyatt put his hand on his mouth as his eyes squinted in an attempt to detour an emotional display.

Amanda sighed as she smiled at the unusual reaction of this future grandpapa.

Ben spoke, "Whew...well then...what about a name? Benjamin has a great ring to it, don't you think? Or, Benoit if it's a girl?"

They all laughed.

Ben responded, "Hey, that's not funny! Seriously, what do you guys think? Since you got Amanda, Bro, in all fairness, I should be able to name the baby?"

They all laughed again, and this time Ben joined in. "I guess I don't get a consolation prize, huh?" Ben shook his head.

"Oh Ben, you are going to spoil her, I'm sure."

"Her?" Drake asked.

"Yup, it's a girl." Amanda smiled.

"Benoit!" Ben smiled.

Drake spoke, "No offence, brother...but I think that the name "Amanda" is perfect."

The three men nodded at Drake's precision, leaving Ben defeated. Ben smiled, "you win, well played." Drake smiled nodding, "She'll be little Amanda or Amanda junior?"

Amanda responded, "No, no, no...That would be awful. I can see maybe Amy, but not little or junior. Amy sounds so cute...hmm...I like it!"

Drake smiled, "I like it too! Amy it is."

"Amy Dominique Caron." Amanda announced.

All four men in unison said, "Perfect!" surprised at the coincidence, they all raised their eyebrows and looked at one another. Amanda laughed. "I love you goof balls!"

Drake announced, "To our Amy! Where's the champagne?"

CHAPTER 58: CONGRATULATIONS IT'S A CARON

The seed contains information necessary to be the tree. A baby contains information necessary to make choices. Plant a seed and nourish... you get the tree. Have a baby and nourish... you establish the world.

C. A. Solis

It was far from an easy pregnancy. Amanda lost a dangerous amount of weight in the beginning and was put on a special diet, but she had no appetite. She was then put in the hospital under bed rest for the remaining two months, as she refused to have the fetus removed in order that it could grow in the artificial liquidation facility. Women no longer needed to give birth as the fetus could be removed and grown artificially. So far, this means was very successful with a hundred percent survival rates, but Amanda would have none of it. She wanted the full experience of being a mother, and it was as if nature was more than willing to give it to her. She was however quite stubborn and determined to go through the course of it. The bed rest worked and she began to gain weight along with baby Amy.

When the baby first moved, Drake thought that something was wrong because Amanda couldn't contain her excitement. He couldn't see anything at first, but eventually as Amy grew larger, Drake cried with happiness when he noticed Amanda's stomach stretching by what looked like a tiny foot pressing against the walls. He had tried to grab it on a few occasions, but Amy wouldn't play. Each time, she'd stop and not move at all for several minutes. Drake would enter a long period of worry, until he once again noticed a little movement. Drake decided that it would be best if he would instead lay next to Amanda quietly and watch.

Amy was very active in her mother's womb, and visitors flowed in and out of Amanda's room. During the day although day was relative in the Domesticities, Amanda didn't get much sleep, but she liked it that way. She was being forced to keep still on a bed, and that was far from her personality. The visitors were always welcomed, and Amanda's room was filled over and over again with gifts for her and the baby.

The nurses would complain to Ben and Ben would complain to Drake. Drake would try to convince Amanda to get more rest, but she was daunting, so all four men would hall off the presents and turn off the holographic messages every evening. Amanda could fall asleep in a matter of seconds. That always astounded Drake. By the time they'd gather the gifts, she was snoring. She always denied that she snored, but now that there were several witnesses, she defiantly blamed her surroundings.

Drake grew a little worried over her change in personality due to the hormones. He'd never seen her in such a mental state. Wyatt laughed when Drake told him that they may need The Mother to cure her disorder when this was over. Drake laughed after he added that he expected Amy to tear through Amanda's stomach growling. He surprised himself over that hyperbole, and shook his head. "Wow, this has been a wild ride for all of us." Drake chuckled. Wyatt, Drake, Ben and myself (yup, I'm Leo) anticipated the day that Amanda would be back to her normal self. Ben jested that if Amanda refused eventual hormonal regulation...he was going to get her to sit on a nano patch. We all laughed.

The day finally arrived. Amy was due on the 8th of April, but was born on the 10th. Those two days drove us all off the edge. It was hilarious to see the famous surgeon Doctor Benjamin Mitchell pace and plead with Amanda to induce the delivery. Of course the baby was not under duress...we were. When Amy came into this world, all four of us were there during delivery. Ben delivered Amy, cleaning her and Amanda then he suddenly ran out the door! Drake ran after him. Amanda laughed and laughed, while the nanopatches moved in to stitch any tearing.

Wyatt walked up to Amanda and kissed her forehead, "You're amazing, Dart, amazing." He smiled.

"Ah, wuz nuttin but a little ting." Amanda smiled.

Drake walked back in with Amy in his arms. "She's a perfect little ting, ain't she?" He placed her in Amanda's arms.

Wyatt brushed his index finger over Amy's cheek. "She's a little jewel, a perfect little jewel."

"She's a Caron, Dad. Her Grandpapa is the world's treasure. I think you're on to a great nickname." Amanda smiled then with a look of concern, she looked at the door. "Where's Ben?"

Drake responded, "I had to kill him."

Amanda smiled, "Uh huh, that dang nappin' kidnapper deserved it."

Ben walked in, "Hey, I was just about to score and spike the football, but Drake tackled me. I guess my team lost. Just figures."

Amanda gestured for Ben, and he moved closer and then next to her. She spoke softly, "This is your Uncle Ben...one of the finest people that you will ever know."

Ben's expressions softened, and he kissed Amanda and Amy on the head. "You did good, Amanda, real good."

It was my turn, I walked up to Amanda's side and she took my hand, "Here's Uncle Leo. Come to think of it Leo, I had the strangest but most wonderful dream about you yesterday." "Oh?" I said and smiled. She continued, "You were telling me that everything would go well and I wasn't to worry. I thought that maybe it was The Mother...but anyway, it was one of the most vivid dreams that I've ever had. Apparently my subconscious loves you very much." Amanda grinned.

"And you?" I asked.

"Oh, so very much, Leo. I'm so glad that you're the one doing our biographies. I couldn't imagine a better perspective than yours. You're quite extraordinary in that regard."

"Thank you" I said, softly adding, "I try as best I can in this form." Perhaps they didn't hear...

CHAPTER 59: THE AGE OF TECH

Writing is the continuation of politics by other means.

Philippe Sollers

The Carron's not only continued on in technological advancements but in raising Amy. She appeared to be their proudest achievement.

Wyatt pondered the need for technology. If only our human brain was in fact an integrated awareness, as perhaps The Mother was...we may not have need for technology.

Technology was mostly in place because the human brain was too slow a processor for our needs. We were evolving in intelligence. Advancing now because of our connections to technologies and technology's connections to ourselves.

The brain had amazing ability to abstractly think and problem solve and innovate. Imagination was something that our artificial intelligences seemed incapable. The human brain however was slow. Too slow for what we seemed to need. It took too long to retrieve our memories, such as that of an answer to an exam question, in which only came to us far after we turned in our work. It was affected by chemical and hormonal imbalances. It could be damaged by a bump, or confused by aging. So, we innovated and created machines that were by human measure extremely fast, in order to make up for our brain's inabilities.

Boredom was a healthy motivator in this time of history. The more technological advancements were achieved, the more we demanded of this industry. We seemed to be in a natural evolutionary progression toward technology.

As Proxies we wanted increasingly more challenging scenarios. The distanced needs of when we had been human. These sorts of programs required human programmers because the human brain could in fact imagine infinitely. With the help of The Mother, it was nearly impossible to tell the virtual worlds from this world and some were quite bizarre.

Proxies preferred to remain in virtual worlds in which they could step over the border of ethics. The eventual synthetic brain of the Proxy was easier to put into virtual, and easier to remove. Once in virtual worlds however, Proxies spent most of their time there. They didn't require resources other than electro-magnetic recharge, and the occasional physical which resembled more of a tune up. The Unified Coalition required them to earn their own altruistic points. Proxies aided in any and all fields. Mostly until they earned enough to be hooked back into virtual for months and sometimes years.

They were magnificently alluring. Many were made to look exactly as they did when they were human. But even if you chose to appear as yourself, you were perfected once Proxied. You could choose from an array of what humans considered the most attractive features of the human body. No blemishes or wrinkles, no aging whatsoever other than to gain knowledge.

Anyone could select a Proxy and whether to Proxy at any time. Most had Proxies in place in case of accidents and injuries to their bodies to which transplants might fail toward the continuance of life. The majority of humans chose to remain human unless it was absolutely necessary to Proxy.

Wyatt saw Proxies as our means toward evolution of the human mind. Proxies were incredibly intelligent. If what we mean by intelligence is the ability of save and retrieve information rapidly. They were as fast as our fastest computers. Therefore they found interest in mostly challenging human's imaginations, as they appeared to be missing most of this trait other than to be able to apply quantum looping or the ability to use probabilities in the present to fill in the gaps of the future. Still, this was not comparable to imagination.

The Unified Coalition worried that what we considered the human observer, wasn't quite caught in the Proxies. Something was missing. Humans were in constant change, and it seemed impossible to catch the actual self of the human...the self that knew that its brain was slow, so it innovated technologies. The self that knew when its body was dealing with chemical imbalances. The self that argued with its other selves, realizing that its slow brain wasn't sufficiently deciding its choice. That self was perhaps what couldn't be caught. Maybe it was there in its Proxy, the UC didn't know, no one did. But the Proxies certainly didn't innovate. They instead used the available information that they acquired.

I should perhaps make this clear, Proxies were not zombies. They were extremely attractive and could not be discerned by physical features from human beings. They did however appear detached, although they could talk about any subject. They seemed to become bored relatively quickly, almost as if they were intellectual snobs towards those fully human. Family members became frequently frustrated with their loved ones, as did friends.

Wyatt pondered whether we had actually conquered death with the Proxies. Family and friends mourned those that proxied; even though death had been thought to be detoured. Something seemed to be missing, and Wyatt needed to know what that was.

CHAPTER 60: THE HOLIDAYS

The holidays once brought stress to many...will they continue?

C. A. Solis

It was approaching the Holidays. Wyatt had his Biodome bubble set to snow, and his small lake to freeze over enough to do a little ice skating if wished. He didn't favor winter, even when he was younger, perhaps because he wasn't much of an athlete, and winter sports required some amount of balance of which he lacked. Ben along with The Mother's assistance had performed successful corrective surgery, and Wyatt was able to skate, still, Wyatt's Biodome was set to bring back warm weather as soon as the holidays ended.

The modest boat at the end of his pier would once again begin to rock on the small warm waves; and the rhythm and sound of the rope being pulled and loosened would once again calm Wyatt as he sat on the beach under whatever sky he chose.

Sometimes he would select from the various storms happening around the planet and watch the lightning from the safe location under his bio domed sky. He'd still jolt on occasion when the sky lit up and crackled as the storm reached its peak rumbling louder and louder. It was better than fireworks, and the warm sand felt great against his back as he looked up at the show. Lately as the storm moved away making the rumbling softer as it departed he had fallen asleep a few times, only to have Doctor Feynman wake him from his naps, as he had requested. He thought that maybe his age was finally becoming a factor to his health as his short naps increased. He was approaching two centuries in age. For these times however, Wyatt was considered middle aged. The human was now expected to live to an average of four hundred years.

The Caron's mansion sat in the middle of a massive biodomed bubble. It had been renovated by Dominique to fit into the bubble, and offered Wyatt and herself views from several windows throughout the home. The top floor however was reserved to be Wyatt's laboratory, and it was surrounded by windows. Wyatt needed nature's beauty in order to concentrate, and he required the choice to be out there amongst nature. His choice to be indoors doing his research made him focus better, so that the beach and lake were a necessity if Wyatt was to continue to contribute through his innovative brilliance. Dominique had made sure to accommodate Wyatt in nearly every possible way.

Wyatt had just been awakened on one of his laboratory couches by Doctor Feynman. In Doctor Feynman's distinctive Bronx accent and typical hand gestures, "Eh Doc, Doc Caron, it's time you got up, your grand-daughtah is heah." Wyatt yawned, "Oh, thank you, Richard." he yawned again. "How long was I sleeping? "Sixty three minutes." "Thanks, that will be all for now." Doctor Feynman nodded and disappeared. The Mother's holographic abilities never stopped amazing Wyatt.

"Papa! Papa!" Amy ran up the steps to his lab. The hover slide was installed next to the steps, but Amy preferred to challenge her endurance on the long and wide stairway. By the time she reached the top, she was panting. "Papa, papa! Where are you?"

Wyatt opened the door to the lab. "What is it, Jewel?"

"Did you see all the presents?!" She smiled largely.

"Of course I did. I'm the one that put them there."

"Wow! There's more than last year!"

"I know, I know. I went a little over board." Wyatt chuckled. "But wait until you see what your Uncle Ben has...I'm sure that he beat your Papa this year."

"Really? Oooh, hmm, I asked him for the new design model of the Titan Biodome. It's small and elusive and yet so resourceful. You know, Papa, it's cutting edge space technology. The walls are the thinnest ever, and they incorporate not only glowing photonics, but the virtual realities are enough to keep us humans entertained indefinitely. The design models are said to be close to the real thing. Really, really, really amazing. Uncle Ben was talking to Uncle Leo about it, and Uncle Leo knows one of the scientists that's working there...and...hmm... I wonder! What do you think, papa?" Amy's anticipation was delightful to Wyatt.

Wyatt laughed, "You remind me so much of your grandmamma. She would be beyond thrilled to see her granddaughter follow in her footsteps."

"Difference is that I want to design for space and the terraforming missions. I want to experience the endless abyss, and build into it leaving stepping stones wherever I can." Amy looked toward Wyatt for approval.

"Yup, yup, yup and that you shall." Wyatt said as he escorted Amy to sit down on one of the computer station chairs.

Amy smiled. Her large green eyes sparkled through her obsidian bangs.

Wyatt questioned as he took a seat, "That reminds me, how are things going in your graduate classes?"

"Very well, Papa. Mom and Dad will probably tell you all about it. I love to learn." Amy stated with affirmation.

"That no doubt runs in this family, Jewel. You're going to be an incredible astronaut."

Amy smiled largely at the prospect.

"Well, we should head downstairs so you can open some gifts. Uncle Ben should be getting off of work shortly and probably is expecting you?"

"Yeah, Mom and Dad were talking to him before we left home, and he told us to stop by after we visited with you. But Dad has some work to finish, so Mom told Uncle Ben that she and I would be there. And now...I am soooo excited! This is going to be the best holiday ever!"

"Then, let's get the ball rolling!" Wyatt announced as they stood to descend the steps together.

"Race you!" Amy ran down the steps.

"Don't you dare open anything until I'm there, Jewel?!"

"I won't!" Amy screeched.

Wyatt shook his head in laughter. She amazed him every moment.

Before Wyatt entered the living area in which Amy and Amanda waited, he instructed one of his AIOs to start the open interface in which Drake could watch the events unfold from his home.

Each year, Wyatt would arrange to have several educationally challenging games selected and wrapped for Amy. He knew that she would learn them very quickly and then have them recycled. It was therefore a task to those programmers at WCIOQT to come up with new and innovative constructs. He was assured that they were in fact working on next year's possibilities.

With each of the games that Amy unwrapped came a tearful hug and thank you to her Papa. Wyatt smiled with true happiness.

Drake stated from his home laboratory, "Those look incredible! You and Daddy will have to see how fast we can beat them, huh?"

Amy looked at her father's holograph, "Did I just hear a mouse? Weeee, Dad?"

"I helped last year! Well, a little?" Drake smiled.

"You're smart, Daddy...real smart, but sometimes you just have to give in to your intuitions. Take some time off. Don't let stuff get to you."

"Yeah, you might be right, Baby Girl. I admit that those things can be frustrating!"

"I'll learn ya one of these days, Daddy...you just gotta follow my lead. If I can conquer Papa's games, I might just be able to make life sustainable anywhere."

Amy looked at her grandfather, "we're about to set up another small Domesticity on Enceladus, Papa, which will allow more scientists to reside and research. The programmers here use The Mother in order to challenge in the most realistic way, in order to train astronauts."

Wyatt cleared his throat, "Open this one, Jewell." Wyatt handed Amy an envelope.

Amy didn't hesitate, she quickly slid her finger over the security strip and the envelope popped open. She pulled out what appeared to be a ticket, and read the inscription. Her eyes opened widely. "It's a ticket to be trained for a mission to Enceladus! No way! Seriously, Papa?"

Wyatt smiled, "Only if you want to go?"

Amy jumped up off the floor, and threw her arms around her grandfather, hugging him tightly.

Wyatt smiled largely, "I'm glad that you like it."

"Like it?" Amy peered into her grandfather's eyes. "Fly me to Saturn!" She laughed with delight.

Drake's response was overshadowed by his own thoughts, "We'll do just that, when we have a little more time, Baby Girl." Drake bit on his lip as his focus veered toward his project.

"Huh?" Amy realized that her father's attention was being captured by his work.

"I love you, Daddy. Don't worry, ain't nuttin but a little ting." Amy blew him a kiss.

Drake nodded smiling slightly at Amy engaging in her mother's euphemism in regard to Amy, and blew a kiss back. Drake appeared embarrassed. "Oh heck, there I go again. Sorry Sweetheart."

"Don't worry, Daddy...I understand."

Drake nodded then grinned, "However, I'm closer to Uncle Ben's dome, so, this year I'll be able to beat you and your mama...brouhahaaa!

Wyatt hugged Amy, and she snuggled into his chest, smiling. "Daddy, you're the best." Drake smiled largely as his image faded. "So are you, Papa...so are you."

Wyatt contently smiled, "I had to be in order to have a granddaughter as amazing as you are." Wyatt slightly tightened his hug. "So, I wonder if any of these games will last long. That chip implantation to The Mother has sure been a success. Your generation's intelligence is truly remarkable."

"You think, Papa? We still don't know what's out there, not really. Are there other Universes? What's the abyss, anti-photons, really?"

"We're relatively sure that there are other Universes because of inflation, Amy. That was proven over a hundred years ago? We also know that light can be produced from darkness by using spinning mirrors."

"Oh, I know all that, Papa. We've detected gravitational waves, but we still don't know from where. We should be so much further on the Kardeshev scale. We should be close to building other Universes." Amy giggled. "But seriously, that's why I'd love to terraform other locations out there in order that we can learn more and maybe breakthrough more and more barriers. We can travel safely through our holograph rooms, and see through the eyes of our space probes, but we don't get our hands dirty. Know what I mean?"

"Hands dirty? Sounds to me like one of your Mother's many euphemisms?"

Amy smiled looking up at her grandfather's face. "She says that you like to get your hands dirty?"

Wyatt kissed the top of her head. Realizing that his granddaughter was not only following in her grandmother's footsteps, he sighed and contently smiled.

CHAPTER 61: LEAVING ON A HIGHWAY

While there's life, there's hope.

Marcus Tullius Cicero

"Okay, Amy...time to leave."

Amy's eyes widened. "Uncle Ben's?"

Amanda nodded, "Uh huh, say goodbye to Papa and let's go."

"Yes!" Amy shouted, then looked at her grandfather.

"Oh, sorry Papa. No offense...we'll be back before you know it." Amy smiled.

"No offense taken, Jewel. I hope it's what you're hoping for this year."

"Whatever it is, is fine with me, Papa. He can't beat you this year!" She grinned. "Uncle Ben is the best. Uh, best Uncle...and you're the best Papa, Papa." She squeezed Wyatt's waist in the strongest hug she was capable.

"Whoa Jewel, you'll be back soon." Wyatt smiled kissing her on her head again.

"Driving?" Wyatt asked Amanda.

"Yup, I drove here from our place, the manual lane was clear." Amanda replied.

"Yeah, a lot of people are choosing the auto driver. Faster and safer, but I guess we like to get our hands dirty on occasion." Wyatt grinned.

Amanda smiled and nodded, "In the car young lady."

Wyatt's security AIO 563 opened the exit door. Wyatt spoke, "Thank you, five six three."

"Are you ever going to give it a name, Dad?" Amanda asked.

"Nope, not after Juan. I actually miss him...so, nope, never again." Wyatt announced.

"How are the repairs coming along?" Questioned Amanda.

"Well, your Grandpa put some really strong protective programs into Juan's system, so it was nearly impossible to bring him back to his original programming They updated him, but it just wasn't the same Juan, so I decided to let him go." Wyatt appeared sad.

"I, I'm sorry, Dad. It's like a death isn't it?" Asked Amanda.

"It was in a way, I guess, Dart. We do tend to anthropomorphize."

"Yeah, I remember doing that to a doll once. When her arm fell off, I cried. Thinking back, it was kind of creepy." Amanda grinned.

Wyatt laughed. "Well, you two better be off. See you later." Wyatt hugged Amanda and then Amy. They both waved as they entered the vehicle. Amanda could be heard saying "auto off" as she took the steering wheel and headed toward the manual operation lane in the connective tunnel.

Wyatt smiled thinking to himself how independent the Caron's were in regard to their choices.

CHAPTER 62: THE FINAL CHAPTER OF OUR LIVES

They rushed them in the ambulance and as they took their last glance, they smiled and looked up at me, "Don't worry, we'll be alright, you'll see." They closed their eyes. My heart broke, my soul cried.

C. A. Solis

The scene was horrific. Two vehicles twisted into what looked like an abstract art form. Ambulances were there within three minutes, having been driven by AIO. The AIO medics quickly jumped out of their ambulance and began cutting into the nanotubular constructs of the vehicles with special tools specifically designed to be able to cut into such materials. They eventually managed to remove the bodies. Two were barely alive. Neither of the three victims had activated their shields, indicating that it was sudden and unexpected.

The dead, a young lady named Sarah Hutchinson. Her fiancé's holographic message played over and over again at the site of the accident. "I love you, Sarah. See you soon." It could only be guessed that her attention to driving had been removed briefly but enough to detour from her lane.

Speeds of both vehicles were estimated to be well above one hundred miles per hour.

The two survivors were immediately put into the ambulance's suspension chambers. The human could remain alive for several years in such chambers. These were the same chambers that our astronauts originally used in their travels within our solar system (it was easier and safer to experience outer space via probes). However, if the body were severely damaged, these chambers simply kept them in their current condition. It was therefore imperative to correct such damage.

Drake was notified by his holograph of Doctor Einstein, and he arrived by AIO operated vehicle within five minutes. He told his vehicle to meet him at the hospital, and quickly boarded the ambulance with his wife and daughter. He sat between them, rubbing the translucent tubes and repeatedly telling them to hang on.

The holograph of Doctor Richard Feynman notified Wyatt, as I waited at the hospital. I had decided to keep some distance because of the current chaotic ongoing. I would continue to write my observations unnoticed.

Ben was called back to work at the hospital. The Administrator notified him of the situation. He met the ambulance as the AIOs moved the suspension chambers. The chambers hovered and smoothly glided to the rooms made available for transplantations and transfers if necessary.

Ben's eyes met Drake's. Ben hadn't time to put on his scrubs correctly, and the blue shirt hung down past his shoulders. He quickly checked the chamber's readout and decided to work on both immediately. "Apply the jelly! Put them in virtual worlds! We need to move now!" Ben screamed at Taylor (his favored nurse) and the other surgeon.

 Nano jelly was applied to both females in order to allow the antibiotic and anti-viral nanos to protect against any possible infection. Amanda's and Amy's rooms suddenly became holographic projections of their internal organs, and Ben walked through Amanda's. Drake walked and conversed with the information, fed by The Mother, of Amy's organs. The Mother returned discussions via the hospital's computer voice capability.

 Several other surgeons at the hospital wanted to help. Ben called for Taylor and another trusted physician, and told a couple of the others to monitor and follow-up. Surgeons moved in and out of the rooms and their image relays. A few surgeons took areas of the images and had them sent to several 3D copiers which would eventually allow them to view the complications and compare the damages and corrective procedures before the actual grown organs arrived.

 The Unified Coalition had been notified through The Mother. Speaker Jaiobian called the hospital and informed them that whatever the UC could do, they would do. Security was sent to the hospital. Once security was established, information was sent via The Mother. Thousands began to gather around mostly out of curiosity.

 Wyatt arrived through a barrage of spectators and screaming questioners. Security moved through the crowd and escorted Wyatt into the hospital.

 Drake knew medical applications, but he was rusty at best. He went from one room to the other through the connective door between both. He walked into and around the large holographs of his daughter and wife's internal conditions. Their hearts were beating rapidly which was normal considering the extreme trauma. He grew more and more frustrated and demanded that The Mother do something. The Mother was silent, except to relay the deteriorating circumstance. Amanda and Amy would be required to request The Mother. Drake could not override their free will. Such was the result of the limitations set for the quantum computer.

 Ben manipulated the nanobots as they closed off the bleeding toward some of the surgeries. The machines moved quickly as they attempted to patch the damage. "When are their organs getting here?! Use the damn printer if you have to, but I need those organs!" Unfortunately the printer could take hours to produce the amount of parts necessary, which were needed now. Besides, the copies would be synthetic based, and these ladies were human and their bodies might possibly reject the attempts. Ben knew this as his frustration increased. Various parts however had been produced, and the other surgeon as well as Ben had been working vigorously to replace some of the damaged tissue and muscles while waiting on the major organ replacements.

 Within a few minutes many other surgeons were arriving. All were prepared to stand-by and relieve Ben and the other surgeons when necessary.

 Taylor walked up to Drake and applied a nanopatch. "What?" Drake looked at Taylor.

"It's just a sedative, Doctor." Taylor looked into his eyes sternly.

Drake nodded. The nanos instantly took effect.

"Besides, Doctor Caron, it's a very small dosage. Doctor Mitchell will have to give you another here shortly."

Drake looked toward Ben, and began to ramble, "They, they, they're everything, Ben. Life has no meaning, has no meaning without them Bro, you have to save them. You have to save them! Mother, mother...save them." His voice slightly faded off.

"That's exactly what we're going to do, Brother. Hold it together, okay? The bots are working so far, and their organs should be here shortly. I've been told that Wyatt and Leo are on their way here as well. The best minds in the medical field."

Drake knew better. He could see that the machines were failing.

Note from Leo: I have continued this biography in part two, Book II, of the trilogy, titled; DEATH'S DEATH.